Welding, Joining and Additive Manufacturing: Experiments, Materials and Modelling

Welding, Joining and Additive Manufacturing: Experiments, Materials and Modelling

Guest Editors

Swarup Bag
Christ Prakash Paul

MDPI

Basel • Beijing • Wuhan • Barcelona • Belgrade • Novi Sad • Cluj • Manchester

Guest Editors

Swarup Bag
Department of Mechanical Engineering
IIT Guwahati
Guwahati
India

Christ Prakash Paul
Additive Manufacturing Technology
RRCAT Indore
Indore
India

Editorial Office
MDPI AG
Grosspeteranlage 5
4052 Basel, Switzerland

This is a reprint of the Special Issue, published open access by the journal *Materials* (ISSN 1996-1944), freely accessible at: www.mdpi.com/journal/materials/special_issues/VYRP3B5K70.

For citation purposes, cite each article independently as indicated on the article page online and using the guide below:

Lastname, A.A.; Lastname, B.B. Article Title. *Journal Name* **Year**, *Volume Number*, Page Range.

ISBN 978-3-7258-3256-9 (Hbk)
ISBN 978-3-7258-3255-2 (PDF)
https://doi.org/10.3390/books978-3-7258-3255-2

Contents

About the Editors

Swarup Bag

After completing his PhD at IIT Bombay, India, Dr. Swarup Bag joined the Department of Mechanical Engineering, Indian Institute of Technology Guwahati, as a faculty member in 2011. He currently holds the post of Professor in the Mechanical Engineering department at IIT Guwahati. His primary areas of research are the fundamental process modeling of welding and joining technologies, wire arc additive manufacturing, the optimization of manufacturing processes, and recrystallization in metal forming processes. Dr. Bag has published about 89 international journal papers, 23 book chapters, and 6 applied patents related to the welding and additive manufacturing processes. He is the recipient of the 'The Royal Arc Award 2009' from the Indian Institute of Welding for the best PhD thesis in welding.

Currently, eight students have received doctorate degrees (PhDs), and twenty-seven master projects (M. Tech) have been completed so far under his guidance. Five research projects have been completed so far in his credit. Dr. Bag has developed six MOOCs (NPTEL) and corresponding teaching materials.

Christ Prakash Paul

Dr. C P Paul is leading a research group in the area of Additive Manufacturing Technology at Raja Ramanna Centre for Advanced Technology (RRCAT). He has been recognized as one of the top 2% most Influential scientists worldwide consecutively for the last five years (i.e., 2018 to 2023) in the list prepared by Stanford University, US, and published by Elsevier B V.

Dr. Paul has 25+ years of experience in the research and development of LAM, from system development to its applications in engineering. His research programs focus on next-generation additive manufacturing systems and processes. He has been instrumental in promoting LAM technology in India through collaborations with other national labs and reputed academic institutes by extending an indigenously developed facility for experiments. He has written chapters in 26 books and published over 150 research papers in peer-reviewed international journals. He authored a textbook on Additive Manufacturing for Undergraduate/Postgraduate Students, Faculty Members and practicing AM Engineers.

Dr. Paul is the Founding Director of AIC π-Hub – Incubation Centre of RRCAT and contributes significantly to the technology translation of RRCAT technologies, expertise, and knowhow for Indian industries and startups.

He earned a Doctoral degree in Mechanical Engineering on the Laser Additive Manufacturing of Engineering Components in 2005 and was a Research Professor at the University of Waterloo, Canada, during 2006–2008.

His extensive experience and leadership have allowed him to address additive manufacturing challenges holistically and promote this strategic research area. His efforts have been recognized with many awards including the highest DAE Excellence in Science, Technology and Engineering award "Homi Bhabha Science & Technology Award".

Preface

Welding and joining with difficult-to-weld materials, dissimilar materials, and materials with complex geometric configurations are always challenging tasks. The extension of welding technology to layer-by-layer deposition brings the revolution of additive manufacturing. Several precision manufacturing technologies in the frame of additive manufacturing or 3D printing are in close proximity to current industrial demand. Additive manufacturing or 3D printing can replace multiple manufacturing systems and produce complex components with great precision in a single manufacturing cycle. Although 3D printing using raw materials in the form of metallic powders produces miniature components with great precision, the production rate and cost of the component are primary concerns. Hence, the industry is moving towards developing wire arc additive manufacturing systems. In addition, welding or additive-manufactured components without any metallurgical issues or process defects and adequate surface finish are critically crucial for industrial product development. Hence, the requirement of the minimum wastage of raw materials, environment-friendly technology, and the optimum utilization of resources is the current demand for the development of a green manufacturing industry. With this anticipated view, the current reprint attempts to provide an elementary presentation of the subject.

It is assumed that the readers have already studied the fundamental aspects of the welding and additive manufacturing processes. Advances in this area are intended to provide students, researchers, academicians, and experimentalists with critical thinking, approaches, and clear ideas about the subject's development. Considerable emphasis is given to constructing the advanced topics and critical issues of welding and additive manufacturing. This edited reprint covers the most important findings for the development of fusion welding, which includes high-energy beam welding, arc welding, ultrasound-assisted fusion welding, and vacuum brazing. The critical aspects of the technological advancement of cold metal transfer for the development of additive manufacturing are the thrust area of this book. The in-process temperature measurement technique for additive manufacturing makes a significant contribution. The processing of both metallic and non-metallic materials, which include low- and high-molecular-weight polymers, titanium alloys, steels, and superalloys, is covered here. The development of technology with the aid of mathematical modelling considerably reduces costly experimental affairs.

Many people have participated directly and indirectly in preparing this edited book. We would like to acknowledge all of them and the contributory authors of individual chapters. We would like to show our gratitude to MDPI for their efforts in allowing us to prepare this reprint on time.

Swarup Bag and Christ Prakash Paul
Guest Editors

materials

MDPI

Editorial

Welding, Joining, and Additive Manufacturing: Experiments, Materials, and Modeling

Swarup Bag [1,*] and C. P. Paul [2]

1 Department of Mechanical Engineering, IIT Guwahati, Guwahati 781039, Assam, India
2 Raja Ramanna Centre for Advanced Technology, Indore 452013, Madhya Pradesh, India; paulcp@rrcat.gov.in
* Correspondence: swarupbag@iitg.ac.in

1. Introduction

With the advancement of technologies, welding and joining have become an integral part of advanced manufacturing systems. The joining of new materials, dissimilar materials, composite materials, different geometric configurations, and miniature components poses significant challenges to the established joining technologies. The industry's primary concern is establishing sustainable joining technologies for non-conventional forms of materials and using these energy sources for developing additive or printing technologies through experiments and mathematical modeling. Although joining metallic materials through deposition is an established process, strategic material transport becomes indispensable to establish a layer-by-layer manufacturing process. Material deposition from the principle of fusion welding in a controlled manner is the backbone for wire arc additive manufacturing (WAAM) technology. However, mathematical modeling of welding and additive manufacturing processes helps to establish the in situ physical behavior of the process. In contrast, simple experiments do not always allow for the reaching of any definite conclusion. It requires an understanding of the phenomenological behavior of thermo-fluid or thermo-mechanical analysis with the application of material science. In addition, the establishment of structure–property relation requires proper characterization techniques of the components, as well as measurement of temperature. Given the anticipated objective, this Special Issue presents several articles in the diverse directions of modeling, characterization, and measurement techniques of welding, joining, and additive manufacturing issues for materials ranging from simplistic to exotic.

check for updates

Received: 7 January 2025
Revised: 16 January 2025
Accepted: 17 January 2025
Published: 19 January 2025

Citation: Bag, S.; Paul, C.P. Welding, Joining, and Additive Manufacturing: Experiments, Materials, and Modeling. *Materials* **2025**, *18*, 445. https://doi.org/10.3390/ma18020445

2. Short Description of the Articles Presented in This Special Issue

The research on welding, joining, and additive manufacturing processes builds up through the coverage of ten individual articles broadly focused on experiments and physics-based models of very critical aspects. This Special Issue mainly focuses on fusion welding and wire arc additive manufacturing (WAAM). The numerical solution of the process starts with the governing equations and boundary conditions using suitable solution techniques. The analytical solution for process stability of fusion welding and additive manufacturing is the most promising input of the present editorial. In addition, the bubble collapsing of fusion welding is invariably an important feature. The heat source model of fusion welding at cryogenic conditions brings new dimensionality over the conventional model. The temperature measurement technique for the additive manufacturing process is an excellent contribution to advancing the current manufacturing system. The mechanical and microstructural characterization of welded and additively manufactured components of advanced materials brings innovative information to the manufacturing community. All these papers have been peer-reviewed for publication in this Special Issue. A short review of all these papers is presented as follows. The first six papers are relevant to welding and joining, and the remaining four papers describe the critical aspects of additive manufacturing.

The paper entitled "Comparative study on the behavior of keyhole in analogy welding and real deep penetration laser welding" depicts the experimental simulation of keyhole formation to understand the mechanism [1]. The argon gas stream is passed through different liquid mediums like water, NaCl solution, and Ga–In–Sn alloy to produce the keyhole, and the analysis is performed from the images of a high-speed camera. The compressive pressure consists of gas jet-induced pressure, surface tension at the interface, and hydrostatic pressure at the liquid surface, which is compared to the vapor pressure of the actual deep penetration laser welding system. The liquid pressure in analogy welding and vapor pressure by evaporation in laser welding are the main driving forces for opening the keyhole, whereas the hydrostatic pressure and surface tension promote the closing of the keyhole.

The paper entitled "Analytical simulation of the microbubble collapsing in a welding fusion pool" shows the ultrasound treatment in the weld pool to influence the solidification behavior and refine the microstructure in a vibration-assisted welding process [2]. The approach lies in finding the optimum position of the working plate to remotely vibrate it such that the ultrasound system will create a cavitation vibration at the weld pool. It is concluded that 20 kHz frequency over a 0.1 s solidifying phase is sufficient time to make the microbubbles burst in the semifluid, which contributes to grain refinement and better shaping of intermetallic phases.

The paper entitled "An improved method for deriving the heat source model for FCAW of 9% Nickel steel for cryogenic tanks" describes the integrated evolutionary optimization algorithm with finite element-based heat transfer analysis [3]. The joining method is adopted here as flux-cored arc welding (FCAW). The temperature is experimentally measured at a series of thermocouple points and is compared with computed data to minimize the error. This inverse analysis is directed toward finding the most suitable double-ellipsoidal heat source parameters. It is concluded that the adjustment of the heat source parameters is befitted with a relatively coarse mesh of reduced computational time. This simplified heat conduction model accounts for a large set of heat source parameters to reach an optimum level with a low computational time.

The paper entitled "Microstructure and mechanical properties of steel and Ni-based superalloy joints for rotors of high-speed electric motors" portrays the mechanical and microstructural characterization of dissimilar joints of magnetic (stainless steel) and non-magnetic (Inconel) laminated sheets [4]. This welding method is applicable for high-speed electric motors where laminated sheets are joined by vacuum braze and hot isostatic pressing (HIP) processes. The results show that both methods are equally efficient. The finite element simulation of stress distribution is performed at the operating speed of the rotor. The minimum shear strength of the joint is estimated, which is close to the experimental results, essentially showing the effectiveness of the developed numerical model.

The paper entitled "Microstructural and performance analysis of TP304H/T22 dissimilar steel welded joints" defines the effectiveness of dissimilar austenitic stainless steel and austenitic/ferritic stainless-steel pipes for high-temperature (550 °C) applications [5]. The weak position of the welded joints is located through thermo-mechanical simulation of tensile specimens. A defect-free environment, devoid of any creep cavities and intergranular cracks, is achieved in the present work. It is concluded that the fracture occurs either in the base material or in the fusion zone of the austenitic stainless steel side at a high-temperature tensile load.

The paper "Assessment of changes in abrasive wear resistance of a welded joint of low-alloy martensitic steel using microabrasion test" illustrates the abrasion wear resistance of a welded joint of low-alloy martensitic steel [6]. The samples are prepared by multi-pass MAG (metal-active gas) welding with a preheat temperature of 50 °C and interpass cooling to 225 °C. The wear test is carried out by the ball-cratering method using abrasive slurry over the weld cross-section. The results indicate that the fusion zone has the least abrasive wear resistance, whereas the heat-affected zone (HAZ) exhibits the variable wear resistance. The resultant microstructure in the fusion zone is acicular ferrite with perlite and tempered martensite, and ferrite along with perlite in HAZ mainly brings the disparity in wear resistance.

The paper entitled "Processing of Haynes® 282® alloy by direct energy deposition with arc and wire" depicts the influence of arc energy input and shielding gas in the cold metal transfer (CMT)-based directed energy deposition process on mechanical and microstructural properties of Ni-based superalloy [7]. The arc energy is varied by changing wire feed speed and travel speed to produce thin wall and multi-track blocks. It is concluded that the risk of crack formation and lack of fusion is more pronounced for multi-track blocks than thin walls; however, it is significantly lower than Inconel 718. The samples produce higher strength and lower elongation at enhanced arc energy, whereas nitrogen-containing shielding gas produces higher hardness and low-impact energy and elongation.

The paper entitled "Optimization of CMT characteristic parameters for swing arc additive manufacturing of AZ91 magnesium alloy based on process stability analysis" illustrates the droplet transfer behavior of AZ91 magnesium alloy using swing arc formation by the CMT system [8]. The stability in metal transfer is identified by analyzing the electrical waveforms and droplet images. The horizontal component of arc force mainly influences the stable droplet transport. The process stability analysis of swing arc deposition is performed using the Vilarinho regularity index for short-circuit transfer metal transfer. It is concluded that the optimum combination of CMT parameters could provide a stable deposition process and an aesthetic appearance of the defect-free deposited layer.

The paper entitled "Measuring the cooling behavior of melt pools in (L-PBF) by pyrometry" shows the measurement of temperature using a pyrometer in a powder bed fusion process [9]. The additive manufacturing for a powder-based process is considered here for laser powder-based fusion (L-PBF). The instrument is calibrated by applying a thermal cycle to a printed track and measuring the temperature from both sides, i.e., from the pyrometer and thermocouple points. The comparison brings the calibrated single-color pyrometer data. A two-color pyrometer is also calibrated to enhance the precision of the results. Finally, the temperature is measured for an actual experiment on a single laser track where a distorted signal for smoke formation is considered. A correlation between cooling duration and process parameters such as laser power, scanning speed, laser focus size, and laser focus offset is established. It is concluded that the cooling rate or cooling duration from time–temperature simulation using the proposed pyrometer is helpful for understanding microstructural evolution.

The paper entitled "New partially water-soluble feedstocks for additive manufacturing of Ti6Al4V parts by material extrusion" describes the development of feedstock systems for Ti6Al4V from eco-friendly solvents such as water-soluble binder systems using fused filament fabrication (FFF) and fused feedstock deposition (FFD) processes [10]. The feasibility of two polar binder systems consisting of the low molecular weight polyethylene glycol and the high-molecular-weight polymers poly (vinylbutyral) or poly (methylmethacrylate) has been investigated. The rheological study indicates that a polar binder system can accommodate a solid load of 60 vol% of Ti6Al4V and reach 96% theoretical density after water debinding and thermal debinding and sintering of metal parts. It is concluded that further densification to 99% is possible to achieve after hot isostatic pressing.

It is evident from the presented papers that research on welding and additive manufacturing includes a wide range of manufacturing processes, materials, measurement, and characterization techniques. The fusion welding processes, such as laser welding, FCAW, MAG welding, gas-metal arc welding (GMAW), ultrasound-assisted fusion welding, and vacuum brazing, convey critical aspects and understanding to practical application. CMT is the most widely used metal transfer mechanism for the development of the wire arc additive manufacturing (WAAM) process. Heat transfer analysis, stress analysis, an analytical solution of jet force, electromagnetic force, ultrasonic vibration, and the application of optimization algorithms enrich the modeling approaches for welding and additive manufacturing processes. This Special Issue introduces the application of different materials, such as Ni-based superalloys, austenitic stainless steels, low-alloyed structural steel, martensitic steel, ferritic stainless steel, titanium alloy (Ti6Al4V), magnesium alloy, low- and high-molecular-weight polymers. A range of techniques like similar welding, dissimilar welding, multipass welding, droplet transport mechanism, and feedstock fabrication are learned from these papers. Several experimental evaluations like temperature measurement, temperature-dependent emissivity calibration, metallography and microstructure of metallic and non-metallic samples, images from high-speed cameras, mechanical properties, wear test of welded components, and recording of waveform current are essential outcomes from the presented papers. Till, there is much more new welding and joining technologies that are evolving day-by-day to meet the modern industry challenges. A much more demanding additive manufacturing technology for a wide variety of materials is also in the development stage, which may bring about a revolution in the industry.

3. Conclusions

The objective of this Special Issue effectively aims to improve the welding, joining, and additive manufacturing technologies; characterization of materials and processed components; and establishment of measurement techniques. It collectively shows how to produce beneficial microstructure and properties, avoid common defects, ensure the weldability and printability of commercial alloys, and address the need for post-processing to develop welding, joining, and additive manufacturing processes. The following conclusive statements are derived from the present editorial work. The vapor pressure in the laser welding process is the main driving force for keyhole formation. The microbubble formation through external vibration modifies the solidification behaviors in the fusion welding process. The inverse approach is the most suitable way to find the uncertain parameters of a computationally efficient heat source model. Vacuum brazing is an efficient method for joining laminated sheets of Inconel and stainless steel. A microstructurally sound dissimilar weld joint between austenitic and austenitic/ferritic steel can be achieved by fusion welding. The fusion zone of low-alloy martensitic steel produces the minimum wear resistance property. Heat input is a more effective method than nitrogen shielding gas to achieve high strength in Ni-based superalloy in the directed energy deposition process. A swing arc in the cold metal transfer process produces stable and defect-free deposition of magnesium alloy. A pyrometer-based temperature measurement technique is more flexible for additive manufacturing processes. The mechanism for the bonding of polymers in a non-conventional way and the joining of non-metallic materials shows the path to developing various printing technologies. However, rapid development of additive manufacturing demands much more research on the subject, which is currently the future of the manufacturing industry. In summary, the collection of articles in this Special Issue provides a foretaste view of the recent progress in specific areas of welding, joining, and additive manufacturing. This publication brings the opportunity for students, scientists, researchers, and entrepreneurs to exercise the subject and gain significant knowledge from it.

Author Contributions: Conceptualization, S.B. and C.P.P.; data curation, S.B.; writing—original draft preparation, S.B.; writing—review and editing, C.P.P.; supervision, C.P.P.; project administration, S.B. All authors have read and agreed to the published version of the manuscript.

Funding: This research received no external funding.

Conflicts of Interest: The authors declare no conflicts of interest.

References

1. Hao, Z.; Chen, H.; Jin, X.; Liu, Z. Comparative study on the behavior of keyhole in analogy welding and real deep penetration laser welding. *Materials* **2022**, *15*, 9001. [CrossRef] [PubMed]
2. Teyeb, A.; Salimi, M.; El Masri, E.; Balachandran, W.; Gan, T.-H. Analytical simulation of the microbubble collapsing in a welding fusion pool. *Materials* **2023**, *16*, 410. [CrossRef] [PubMed]
3. Kim, Y.; Kim, J.; Park, H.; Hong, S.; Pyo, C.; Park, G. An improved method for deriving the heat source model for FCAW of 9% Nickel steel for cryogenic tanks. *Materials* **2023**, *16*, 6647. [CrossRef] [PubMed]
4. Scherman, E.; Sikanen, E.; Yeddu, H.K.; Amraei, M.; Sopanen, J. Microstructure and mechanical properties of steel and Ni-based superalloy joints for rotors of high-speed electric motors. *Materials* **2022**, *15*, 6906. [CrossRef]
5. Sun, J.; Wang, T.; Liu, F.; Zhang, Z.; Chen, Y.; Lin, H.; Liu, H.; Zhao, X.; Cheng, X. Microstructural and performance analysis of TP304H/T22 dissimilar steel welded joints. *Materials* **2023**, *16*, 4474. [CrossRef] [PubMed]
6. Ligier, K.; Napiórkowski, J.; Lemecha, M. Assessment of changes in abrasive wear resistance of a welded joint of low-alloy martensitic steel using microabrasion test. *Materials* **2024**, *17*, 2101. [CrossRef] [PubMed]
7. Zinke, M.; Burger, S.; Jüttner, S. Processing of Haynes® 282® alloy by direct energy deposition with arc and wire. *Materials* **2023**, *16*, 1715. [CrossRef] [PubMed]
8. Zhang, Z.; Shen, J.; Hu, S.; Chen, Y.; Yin, C.; Bu, X. Optimization of CMT characteristic parameters for swing arc additive manufacturing of AZ91 magnesium alloy based on process stability analysis. *Materials* **2023**, *16*, 3236. [CrossRef] [PubMed]
9. Pfaff, A.; Schäffer, S.; Jäcklein, M.; Balle, F. Measuring the cooling behavior of melt pools in L-PBF by pyrometry. *Materials* **2023**, *16*, 3647. [CrossRef] [PubMed]
10. Eickhoff, R.; Antusch, S.; Nötzel, D.; Hanemann, T. New partially water-soluble feedstocks for additive manufacturing of Ti6Al4V parts by material extrusion. *Materials* **2023**, *16*, 3162. [CrossRef] [PubMed]

materials

MDPI

Article

Comparative Study on the Behavior of Keyhole in Analogy Welding and Real Deep Penetration Laser Welding

Zhongjia Hao, Huiyang Chen, Xiangzhong Jin * and Zuguo Liu

State Key Laboratory of Advanced Design and Manufacturing for Vehicle Body, Hunan University, Changsha 410082, China
* Correspondence: jxz2014@hnu.edu.cn

Abstract: In deep penetration laser welding, the behavior of the keyhole has an important influence on the welding quality. As it is difficult to directly observe the keyhole and detect the pressure inside the keyhole during metal laser welding, theoretical analysis and numerical simulation methods are commonly used methods in studying keyhole behavior. However, these methods cannot provide direct real information on keyhole behavior. In this paper, a method of analogy welding is proposed, in which high speed gas is used to blow the liquid to generate the keyhole. Relevant process experiments were conducted to explore keyhole behavior in analogy welding and real deep penetration laser welding. The pressure balance of the keyhole, both in analogy welding and real deep penetration laser welding, were analyzed. The laws obtained in analogy welding and real deep penetration laser welding are similar, which indicates that studying keyhole formation and the maintenance principle using the analogy welding method proposed in this paper may be helpful for deep understanding of the keyhole formation and maintenance mechanisms in real deep penetration laser welding.

Keywords: analogy; deep penetration laser welding; keyhole; water model; sandwich method

Citation: Hao, Z.; Chen, H.; Jin, X.; Liu, Z. Comparative Study on the Behavior of Keyhole in Analogy Welding and Real Deep Penetration Laser Welding. *Materials* **2022**, *15*, 9001. https://doi.org/10.3390/ma15249001

Academic Editors: Antonio Riveiro and Ren-Kae Shiue

Received: 20 October 2022
Accepted: 14 December 2022
Published: 16 December 2022

1. Introduction

As an advanced materials processing technology, laser welding has been used in a wide range of scientific and industrial applications, e.g., steel, nuclear power, aerospace, rail transit, automotive, electronics and other major civil or military projects. It plays an increasingly important role in lightweight transportation vehicles by the welding of thin-wall components, Al–Mg alloy and other light materials. Metal materials, such as aluminum alloy and steel, are widely used in the industrial field, and they are the main objects of laser welding. Generally, the modes of laser welding are mainly conduction mode and keyhole mode. When the laser energy is insufficient to evaporate the material, that is, the maximum temperature does not exceed the boiling point of the material, this welding mode is called the conduction mode. When the laser energy is enough to evaporate the material, a keyhole appears inside the material, and, thus, this welding mode is called the keyhole mode. Keyhole mode welding, or deep penetration laser welding, has the advantages of high aspect ratio, high utilization of laser energy and high welding efficiency, and it is the main research direction in the welding field at present. For commonly used aluminum alloy, when the laser power is 3.2 kW and the welding speed is 20 mm/s, deep penetration laser welding can be carried out on 5052 aluminum alloy plates with a thickness of 4mm, and the weld surface is bright and smooth. Of course, deep penetration laser welding also has some limitations, such as high requirements for workpiece assembly, high processing costs, and limited welding depth, especially compared with the depth of electron beam welding.

In deep penetration laser welding, the formation of a keyhole indicates that the coupling mode of laser energy and material has changed; that is, the laser energy absorption has changed from material surface absorption to material interior absorption. The formation and maintenance of the keyhole is the key to realizing deep penetration welding. At present, there are still many defects in deep penetration laser welding, such as weld depression, spatter and multi-humps, which hinder the further development of this technology. The occurrence of defects is closely related to the keyhole behavior caused by the melting, evaporation and flow of the molten pool during the welding process. Therefore, it is of great significance to study the formation and maintenance mechanisms of the keyhole in deep penetration laser welding so as to optimize the welding process and improve the weld quality. However, it is difficult to directly observe the complete keyhole shape inside the metal, due to the opacity of metal, and the pressure balance inside the keyhole is even more difficult to detect directly.

1.1. Keyhole Observation and Theoretical Research

Many scholars have made achievements in keyhole observation and theoretical research on energy and mechanics in keyholes. Arata et al. [1] tried to capture the side shape of the keyhole from the vertical direction of the welding speed using X-ray, and directly observed the fluctuation of the keyhole. Dowden et al. [2–6] successfully built theoretical models of the open keyhole and the blind keyhole, studied the energy balance and pressure balance in the keyhole during the formation and maintenance of the keyhole, explored the generation of thermal stress during laser welding of thin plates, determined mathematically that the viscous resistance related to steam movement was one of the factors causing the violent movement of the keyhole and molten pool, and established a mathematical model to describe the axial movement of the molten pool along the keyhole wall. Semak et al. [7,8] took pictures of the keyhole and the molten pool with a high- speed camera. It was calculated that the metal vapor induced by the laser was significantly larger than the surface tension and static pressure of the molten pool, thus promoting the flow deformation of the molten pool to generate the keyhole. When the keyhole reached a certain size, it would gradually close under the action of surface tension and static pressure, and the process showed a certain periodicity. Kim et al. [9] used a high-speed camera to observe the influence of zinc coating on the morphology of the keyhole in the laser welding process. It was found that, due to the existence of the zinc coating, under the strong evaporation of zinc, the bottom of the keyhole was basically open, while the bottom of the keyhole of the ungalvanized steel tended to close. This phenomenon shows that when considering the pressure balance in the keyhole, the influence caused by the evaporation of elements contained in the material needs to be carefully considered. In order to estimate the keyhole depth in the actual welding process, Lankalapalli et al. [10] established a penetration model of a two-dimensional conical keyhole and linked the penetration depth with the incident power. Solana et al. [11] established an axisymmetric model of multiple reflections inside the keyhole, and ensured the model took into account the inverse bremsstrahlung absorption. In Pecharapa's study [12], the phase transformation of materials in the welding process was taken into account, and a relatively good theoretical model was obtained. In the study by Strömbeck [13], the multiple reflection of the laser in the keyhole was described as the self-focusing of the laser welding system, and a model with higher temperature at the bottom of the keyhole was obtained. Fabbro and Chouf [14] investigated the uniform motion of the keyhole along a straight line, considered the multiple reflection and inverse bremsstrahlung absorption inside the keyhole, and linked the drilling rate of the laser beam with the moving speed of the keyhole. Then, a high-speed camera was used to observe the flow of the molten pool, and it was found that the interaction between the steam plume and the molten pool was the reason for the change of flow characteristics [15–17]. In the experiment, the welding speed was changed from slow to fast, and it was found that the inverse of the keyhole depth and the welding speed were almost linear [18].

With in-depth study of the keyhole, researchers can obtain a clear image of the keyhole in the laser welding process. The results obtained from direct observation experiments of the keyhole have become the basis for studying the characteristics of the keyhole. Jin's team used the sandwich method to directly observe the keyhole [19–22], and constructed a three-dimensional multi reflection model of the keyhole. It was found that most of the positions of the rear wall of the keyhole were not irradiated by the laser. In order to maintain the energy balance inside the keyhole, the energy required for the rear wall of the keyhole would be transmitted from the front wall to the rear wall by the molten pool flow [23–25]. By using the sandwich method, Cheng et al. [26,27] found that, compared with Fresnel absorption, the inverse bremsstrahlung absorption of laser energy by keyhole plasma played a major role in absorbing laser energy, and the electron temperature inside the keyhole was uneven and distributed in the radial and depth directions. Li et al. [28] observed that the steam flowing upward and downward inside the keyhole formed a steam vortex after meeting, and the fluctuation of steam flow and pressure were the key factors leading to the fluctuation of the keyhole. Zhang et al. [29] used the modified sandwich method to observe the keyhole, and summarized the formation of the keyhole into three stages: the fast-drilling stage, the slow-drilling stage, and the quasi-steady state stage. They believe that the key factor to make the width of the molten pool enter the quasi-steady state was the balance between the rotation of the vortex and the lateral flow around the keyhole. Others believe that the steam recoil pressure generated by the energy reflected from the front wall of the keyhole to the rear wall is the main driving force for the deformation of the rear wall of the keyhole, and the rear wall collapses due to large surface tension and hydrostatic pressure during the oscillation of the keyhole [30].

1.2. Numerical Simulation of Keyhole

Numerical simulation is used to study the characteristics of keyholes. Wang et al. [31] established a three-dimensional heat source model, composed of a rotating Gaussian volume heat source and a double ellipsoidal heat source, to simulate the keyhole in the laser welding process. The numerical simulation results showed that the eddy currents formed at the top and bottom of the weld pool were conducive to the overall heat transfer. Cho et al. [32] simulated the flow of the molten pool at the initial stage of the keyhole formation, and found that, in the initial stage, the flow direction in the center of the molten pool appeared as axisymmetric oscillation, which was closely related to the recoil pressure, the cooling and the surface tension. Huang et al. [33] theoretically studied the correlations between the surface area, the volume of the keyhole, the welding speed and the surface tension coefficient, and believed that the surface tension controlled the oscillation period of the keyhole. Bedenko et al. [34] conducted one-dimensional simulation research on the dynamics of keyhole plasma during laser welding, and pointed out that the keyhole plasma had a periodic shielding effect on laser radiation, which caused the absorption of laser energy by workpiece materials to alternately attenuate or stop, resulting in the pressure and temperature oscillations. Pang et al. [35] proposed a mathematical model to describe the dynamic coupling behavior of the keyhole and the weld pool, and pointed out that the surface tension had a great influence on the period of keyhole depth oscillation. The oscillation of the plume ejected from the keyhole was closely related to the instability of the keyhole, and the oscillation frequency was the same as the oscillation period of the keyhole in the depth direction. Li et al. [36] simulated the laser welding of aluminum alloy under sub-atmospheric pressure. The numerical results showed that, compared with atmospheric pressure, the keyhole became wider and deeper, and the hump smaller. With the decrease of environmental pressure, the eddy current on the rear wall of the keyhole decreased or even disappeared, which was conducive to improving the stability of deep penetration laser welding and suppressing the generation of defects. Cunningham, Mayi studied the transformation of the welding mode, and found that there was a clear threshold for the sudden change from the conduction mode to the keyhole mode. During the transformation process, a semicircular pit appeared in the weld pool. Its depth and energy balance were determined by the effect of the recoil pressure on the weld pool [37,38]. Zou et al. [39] believed that during the welding process only the front wall of the keyhole was exposed to the laser beam, and the absorbed energy at the rear wall was mainly absorbed by plasma radiation and multiple reflection of the laser. The depth of the keyhole was mainly determined by the drilling behavior caused by the first absorption of laser energy at the front wall of the keyhole. The molten pool flow around the keyhole and the behavior of the keyhole were studied by some scholars [40–42]. The surface tension is considered to be the main driving force for the molten pool flow. The reduction of the size at the entrance of the keyhole leads to the increase of the shear stress of the steam, which accelerates the formation of spatter. The bubbles are generated by the collapse of the front and rear walls of the keyhole. Based on the pressure balance of the keyhole, Huang et al. [43] analyzed the relationship between the steam plume and the keyhole fluctuation, and believed that the change of the total pressure led to the fluctuation of the keyhole size, and the fluctuation of the plume led to the fluctuation of the hydrodynamic pressure in the keyhole.

At present, many scholars have studied the keyhole and generally agree that the formation and maintenance of the keyhole is the result of the combined effect of energy balance and pressure balance inside the keyhole. The study of energy balance in the keyhole is relatively comprehensive. Due to the keyhole being hidden in the weld pool during deep penetration laser welding, and the pressure in the keyhole being difficult to measure, most of scholars analyze and judge the pressure balance result in the keyhole on the basis of the energy balance, or conduct theoretical simulation research according to empirical formulae, so the conclusions obtained may lack the support of actual experimental data. Several experimental observation methods of keyhole research have been summarized in ref. [44]. Numerical simulation research is mainly aimed at the parameters that are difficult to detect in actual experiments, such as temperature field, pressure distribution, molten pool flow, etc., but the final results need to be confirmed by experimental results.

Limited by the existing technical means, it is difficult to directly detect the pressure inside the keyhole during deep penetration laser welding of metals, which results in difficulties in analyzing the influence of the pressure inside the keyhole and the physical properties of the material on the formation and maintenance of the keyhole. During real deep penetration laser welding of metal, a Gaussian laser can be used to irradiate the workpiece, and a keyhole surrounded by molten metal forms inside the metal. The formation and maintenance of the keyhole depends on pressure balance. When a convergent nozzle is used, a gas jet with a Gaussian velocity distribution can form. If the gas jet is directly used to impact on the liquid, under certain conditions, a keyhole can be formed in the liquid. The formation and maintenance of the keyhole are also related to pressure balance. In this paper, the authors attempted to develop an effective experimental method to study the influence of pressure on the keyhole shape by means of producing an analogy keyhole in liquid materials by blowing focused high-speed gas. So, an analogy method of keyhole formation, namely a keyhole formed by a gas jet impinging on liquid, was used to simulate the keyhole in deep penetration laser welding. Both the analogy welding and real deep penetration laser welding were carried out on liquid and a modified sandwich structure, respectively. In the analogy welding, liquid was used to simulate the molten pool, and the gas jet was used to simulate the vapor. In order to be as close as possible to the real molten metal material in welding, several materials were used in the analogy welding, including liquid alloy material at normal temperature. The behaviors of the keyhole in analogy welding and real deep penetration laser welding were intuitively observed, respectively, and the influences of relevant parameters on the behavior of the keyhole explored. If the behavior of the keyhole, such as keyhole shape, was similar in analogy welding and real deep penetration laser welding, it might offer a way to study the relationship between the pressure in the keyhole, together with the physical properties of the material and the keyhole shape, in deep penetration laser welding, instead of directly detecting the pressure in the keyhole.

2. Materials and Methods

2.1. Interaction between Gas Jet and Liquid

The flow formed by gas jetting from the orifice, nozzle or slit is called the gas jet. In this experiment, the fundamental reason for the formation of the keyhole was the interaction between the gas jet and liquid. A keyhole can form when high speed gas impinges on a liquid surface. The surface tension of the liquid is tangent to the liquid surface at the boundary and has the ability to shrink the liquid surface. The surface tension T can be expressed as [45]:

$$T = \sigma L_b \quad (1)$$

where σ is the surface tension coefficient, and L_b is total length of boundary.

Due to the existence of surface tension, when the liquid surface is no longer flat, the additional pressure ΔP on the liquid surface caused by the surface tension can be expressed as [46]:

$$\Delta P = \sigma \left(\frac{1}{R_1} + \frac{1}{R_2} \right) \quad (2)$$

where R_1 and R_2 are the radius of curvature of the liquid surface, which is positive when the liquid surface is convex and negative when the liquid surface is concave.

The comprehensive pressure P on the gas–liquid interface can be expressed as [47,48]:

$$P = P_a + P_d + \Delta P + P_s \quad (3)$$

where P_a is atmospheric pressure, P_d is dynamic pressure generated by gas jet, ΔP is additional pressure caused by surface tension, P_S is hydrostatic pressure at the position of liquid surface.

The study of a gas jet impinging on liquid is common in the metallurgical industry. The diagram of the keyhole formed by the gas jet impinging on the liquid surface is shown in Figure 1. In the figure, a rectangular coordinate system O-xy was established; p_0, T_0 and ρ_0 are, respectively, the pressure, temperature and density in the nozzle, while p_1, T_1 and ρ_1 are, respectively, the pressure, temperature and density at the nozzle outlet. The temperature and density of the jetted gas are T_g and p_g, respectively. The distance between the nozzle outlet and the liquid level is H, the keyhole depth is h_0, the radius of curvature at the depth h on the keyhole wall is R. The normal of a point on the keyhole wall is n, the outlet diameter of the nozzle is D.

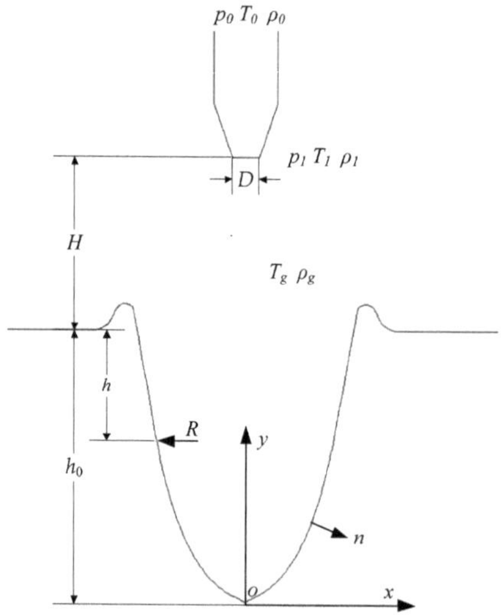

Figure 1. Diagram of the keyhole formed by a gas jet impinging on the liquid surface.

In many studies of the gas jet–liquid system, the keyhole depth h_0, dependent on the jet action, was derived. The balance equation on the keyhole wall can be expressed as [49]:

$$\frac{\rho_g v_g^2}{2} = \rho_L g h_0 + \frac{2\sigma_L}{R_{kh}} \tag{4}$$

where ρ_g is the gas density at the liquid surface, v_g is the velocity of the gas in the jet prior to the collision with the surface of the liquid, ρ_L and σ_L are the density and the surface tension coefficient of the liquid, h_0 is the keyhole depth, R_{kh} is the radius of curvature of the liquid surface at the keyhole depth h_0.

2.2. Pressure Balance in the Keyhole Formed in Laser Welding

In the interaction between the gas jet and the liquid, the hydrodynamic pressure generated by the Marangoni effect and eddy current are ignored. In the following study, it was assumed that the hydrodynamic pressure was ignored, and then the comprehensive pressure P_w on the keyhole, when maintaining dynamic balance in a quasi-steady state, could be expressed as [50]:

$$P_w = P_a + P_v + \Delta P + P_s \tag{5}$$

where P_v is the vapor pressure of the material.

2.3. Experimental Setup

The keyhole behavior was observed by using the experimental setup described in Figure 2 in both analogy welding and real deep penetration laser welding with the sandwich method. In Figure 2a, the experimental setup was mainly composed of a nozzle, an acrylic container, a high-speed camera and corresponding instruments. The acrylic container contained liquid materials. The high-speed camera was placed on the side of the container, and its lens direction was perpendicular to the moving direction of the nozzle. Argon was used as the gas source in the experiment. After passing through the reduction valve and the gas flow controller, the gas inside the gas source was ejected from the nozzle outlet at a certain flow rate, impacting on the surface of the liquid and forming a keyhole therein. In addition, the nozzle was fixed on an ABB Robot to realize the movement. The high-speed camera used in the experiment was a NAC MEMRECAM HX-7S.

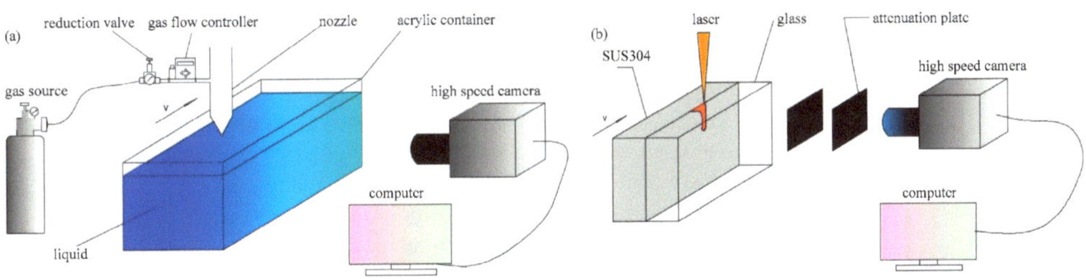

Figure 2. Diagram of the keyhole behavior observation experimental setup: (**a**) Keyhole observation in analogy welding; (**b**) Keyhole observation in real deep penetration laser welding with sandwich method.

In order to analyze the influence of different material properties on keyhole behavior, water, NaCl solution and an alloy of Ga–In–Sn were selected as the liquid materials in the experiment. The viscosity and surface tension coefficients of the materials are shown in Table 1. The variation of surface tension coefficient of NaCl solution with concentration is shown in Figure 3.

Table 1. Viscosity and surface tension coefficient of materials [51–56].

Material	Viscosity (cp)	Surface Tension Coefficient (mN/m)
Water	1	71.4
NaCl solution	1~2	71.4~78.7
Ga-In-Sn alloy	2.4	718

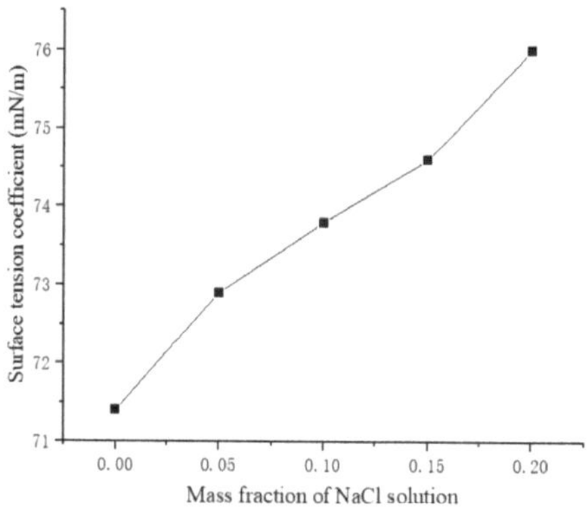

Figure 3. Variation of surface tension coefficient of NaCl solution with mass fraction.

In Figure 2b, the experimental setup was mainly composed of transparent heat-resistant glass, a workpiece, a high-speed camera and corresponding instruments. The glass and workpiece were put together and clamped by a clamp. The high-speed camera was placed on the side of the workpiece, and several attenuation elements were installed on the lens to reduce the light intensity and avoid damaging the photosensitive elements inside the high-speed camera. A fiber laser was employed in the experiment, model IPG YLS-4000-CL. The workpiece was made of SUS 304.

2.4. Procedure

During the experiment in analogy welding, the gas flow was regulated by the gas flow controller. When the initial velocity of the gas jet changed, keyholes with different shapes formed in the liquid. The experiments included the following components:

(1) The liquid surface was suddenly impacted by the gas jet, and the formation process of the keyhole recorded by the high-speed camera. The gas flow changed, and the change of keyhole depth when the nozzle was stationary was recorded.
(2) The nozzle moved horizontally relative to the container at a certain speed v to simulate the laser welding process. The corresponding parameters of analogy welding (moving speed, gas flow rate, distance between nozzle outlet and liquid level, etc.) changed, and the behavior of the formed keyhole was recorded.
(3) Different liquid materials were used, and the above experimental process was repeated for each, and the behavior of the keyhole recorded.
(4) During the experiment in real deep penetration laser welding with the sandwich method, the position of the laser head was adjusted so that the center of the laser beam was located at the interface between the workpiece and the glass. The laser beam was moved at a constant speed along the interface, and the keyhole behavior in welding recorded.

3. Results and Discussion

3.1. Keyhole Formation Process

In this process, the nozzle was stationary, the gas flow rate was 82 mL/min, the diameter of nozzle outlet was 0.2 mm, and the distance between nozzle outlet and liquid level was 5 mm. The adopted high speed camera frame rate was 10,000 fps. The gas velocity V at the nozzle outlet depended on the gas flow rate. The gas velocity V could be expressed as [57]:

$$V = \frac{4q}{\pi D^2} \tag{6}$$

where q is the gas flow rate, and d is the diameter of nozzle outlet.

The formation of the keyhole in analogy welding is shown in Figure 4. It can be seen that under the action of the gas–liquid two-phase flow, at first, a shallow pit formed in the liquid surface, as shown in Figure 4a,b, and then a keyhole was drilled in the downward direction, as shown in Figure 4c–h, and, finally, a certain keyhole depth was reached and maintained, as shown in Figure 4i–l.

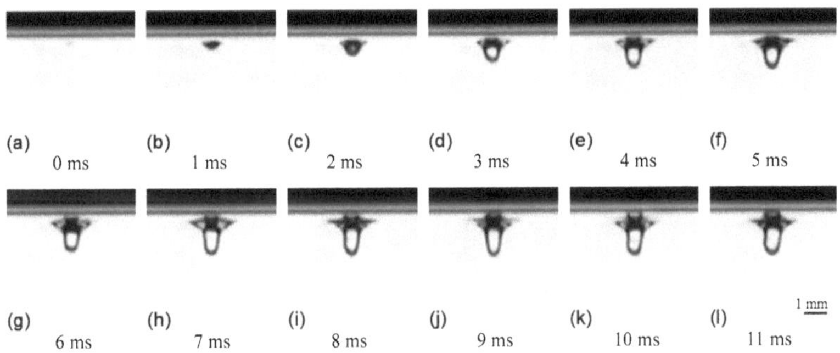

Figure 4. Formation of keyhole in analogy method: (**a**) No keyhole formed at 0 ms; (**b**) Keyhole at 1 ms; (**c**) Keyhole at 2 ms; (**d**) Keyhole at 3 ms; (**e**) Keyhole at 4 ms; (**f**) Keyhole at 5 ms; (**g**) Keyhole at 6 ms; (**h**) Keyhole at 7 ms; (**i**) Keyhole at 8 ms; (**j**) Keyhole at 9 ms; (**k**) Keyhole at 10 ms; (**l**) Keyhole at 11 ms.

When the nozzle was stationary, the variation of keyhole depth with time or gas velocity is shown in Figure 5. Figure 5a shows the variation of keyhole depth with time in Figure 4. It can be seen that during the keyhole formation process, the change of keyhole depth had the characteristics of rapid increase to gradual stabilization, and the keyhole depth would eventually be maintained near a certain value. According to the variation of keyhole depth, the formation process of keyhole in the experiment could be divided into three stages:

(1) Fast-drilling stage.
(2) Slow-drilling stage.
(3) Quasi-steady state stage.

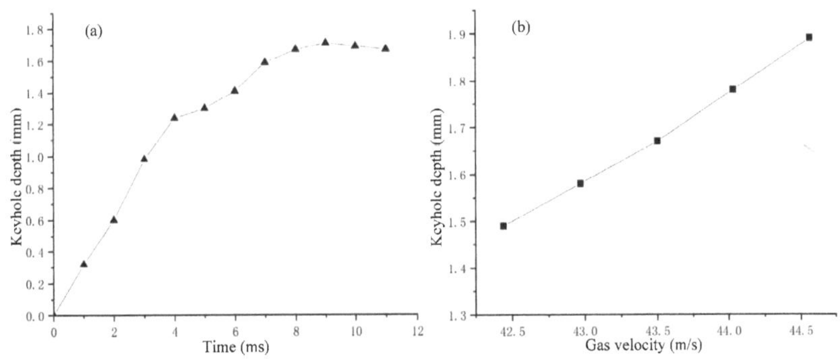

Figure 5. The variation of keyhole depth: (**a**) Keyhole depth with time; (**b**) Keyhole depth with gas velocity.

In the fast-drilling stage, the keyhole depth increased rapidly, and the drilling rate could reach 320 mm/s. When the keyhole depth approached the maximum value, it entered the stage of slow-drilling, at which time the change rate of keyhole depth gradually decreased. Then it entered the quasi-steady state stage, and the keyhole depth was dynamically maintained around a certain value. There were many similarities in behavior between the keyholes observed in the analogy welding and the keyholes obtained in the real laser welding process [25,37,38].

From Figure 5b, it can be seen that the keyhole depth was positively correlated with the gas velocity. In this experiment, the dynamic pressure of the gas jet was the driving force for the formation of the keyhole, which was similar to the effect of the vapor pressure generated by the evaporation of materials in laser welding. Gas flow rate used to analogize laser power may be of significance to the study of laser welding.

At the initial stage of keyhole formation, the dynamic pressure of the gas jet was greater than the surface tension and hydrostatic pressure, so the keyhole depth changed significantly at this stage. With increase of keyhole depth, the gas dynamic pressure directly acting on the liquid surface also decreased, while the hydrostatic pressure increased. Moreover, the curvature radius of the curved surface at the bottom of the keyhole decreased, and the additional pressure caused by the surface tension increased. Therefore, the growth rate of the keyhole depth gradually decreased, and, finally, the keyhole depth oscillated near an equilibrium point.

3.2. *Keyhole in Analogy Welding*

In analogy with laser welding, the corresponding parameters (moving speed, gas flow rate, the distance between nozzle outlet and liquid level, etc.) were considered.

3.2.1. Effect of Moving Speed on Keyhole

The horizontal moving speed of the nozzle was set, while other parameters remained unchanged. The obtained keyhole shapes, at different moving speeds, are shown in Figure 6. It can be seen that when the nozzle moved horizontally, the shape of the keyhole bent to different degrees with the change of the moving speed of the nozzle, and the bending direction was the opposite direction of the moving speed.

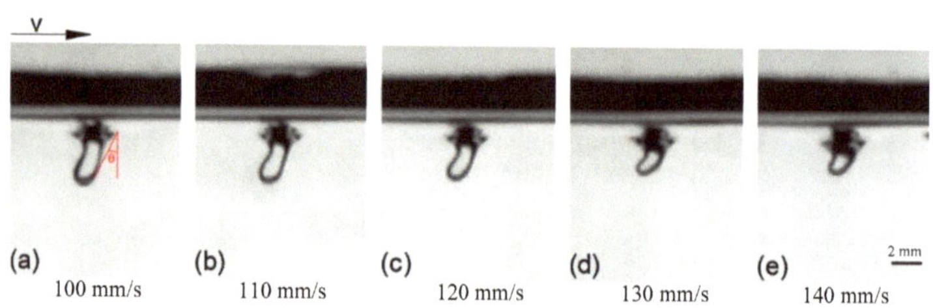

Figure 6. Keyhole during horizontal movement of nozzle: (**a**) Keyhole with nozzle moving speed of 100 m/s; (**b**) Keyhole with nozzle moving speed of 110 m/s; (**c**) Keyhole with nozzle moving speed of 120 m/s; (**d**) Keyhole with nozzle moving speed of 130 m/s; (**e**) Keyhole with nozzle moving speed of 140 m/s.

A bending keyhole is shown in Figure 7a. The angle between the keyhole drilling speed and keyhole moving speed was used to define the bending angle of the keyhole. The tangent value of angle θ was used to characterize the bending degree of the keyhole, and the larger the value, the more curved the keyhole was. The tangent of angle θ was expressed as:

$$\tan\theta = \frac{v_h}{v_d} \tag{7}$$

where v_h and v_d are the moving speed and drilling speed of the keyhole, respectively.

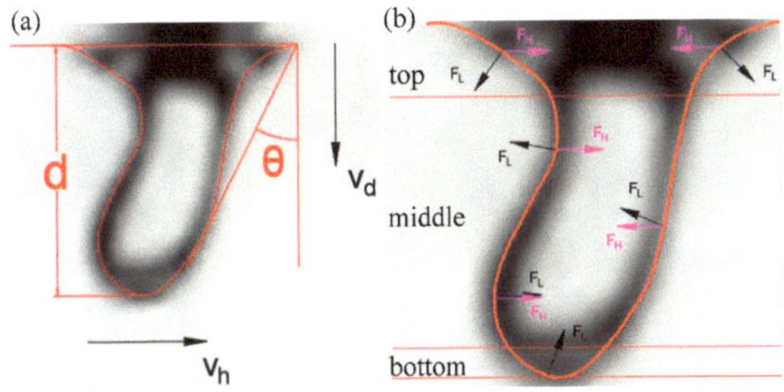

Figure 7. Bending keyhole: (**a**) Angle of bending keyhole; (**b**) Segmentation of bending keyhole.

According to Formula (2), it can be seen that the shape of the keyhole was directly affected by the convexity–concavity of the keyhole wall. In the horizontal direction, the keyhole was always forced to close by the surface tension, while in the vertical direction, due to the change of the keyhole curvature, the effect of the surface tension also changed.

In Figure 7b the bending keyhole was divided into the top part, the middle part and the bottom part.

The part with a smaller radius of curvature at the keyhole inlet is marked as the keyhole top part, the part in the middle of the keyhole, where it is relatively smooth in the vertical direction, is marked as the keyhole middle part, and the concave liquid surface at the bottom of the keyhole is marked as the keyhole bottom part. The resultant force directions of the surface tension in the transverse direction and the longitudinal direction are indicated by F_H and F_L, respectively.

At the keyhole top part, the radius of curvature of the front and rear walls of the keyhole differed, as the surface tension kept the keyhole open in the longitudinal direction, and tended to close the keyhole in the transverse direction. A dynamic balance formed, due to the surface tension, dynamic pressure and hydrostatic pressure. The radius of curvature of the keyhole front wall in the longitudinal direction was small, and the effect of surface tension significant. This was because the gas jet directly acted on the front wall during the keyhole motion, resulting in greater gas dynamic pressure, while the larger radius of curvature of rear wall was related to the flow characteristics of the liquid.

At the keyhole middle part, the front and rear wall differed greatly. From the longitudinal perspective, the direct effect of the dynamic pressure on the front wall was relatively significant, which increased the liquid flow velocity along the wall of the keyhole. The wall had a large radius of curvature and was concave in shape. The longitudinal resultant force of surface tension had a closing effect on the keyhole, but the effect was not obvious. However, a part of a convex shape appeared in the rear wall, as the effect of the surface tension in the longitudinal direction here was the same as that of the gas dynamic pressure to maintain the keyhole opening.

At the keyhole bottom part, the shape was concave, the curvature radius of the keyhole was small, and the effect of surface tension was significant. The gas pressure was the main driving force to maintain the keyhole opening, while the hydrostatic pressure and surface tension tended to close the keyhole.

As shown in Figure 6, it can be seen that, at different moving speeds, the keyhole appeared to have different bends, and so was the keyhole depth. The effect of the change of moving speed on the keyhole depth is shown in Figure 8a. It can be seen that the keyhole depth decreased with increase of the nozzle moving speed, and the relationship between the two was almost linear. The relationship between the bending degree of the keyhole and the moving speed of the nozzle is shown in Figure 8b. The result shows that the bending degree of the keyhole increased with the moving speed of the nozzle, and the relationship between them was also approximately linear. It was obvious that the nozzle moving speed had a linear relationship with the keyhole depth and keyhole bending degree, which indicated that there was a certain correlation between the keyhole depth, keyhole bending degree and the nozzle moving speed. The effect of nozzle moving speed variation in the analogy welding was similar to the effect of welding speed variation in real deep penetration laser welding.

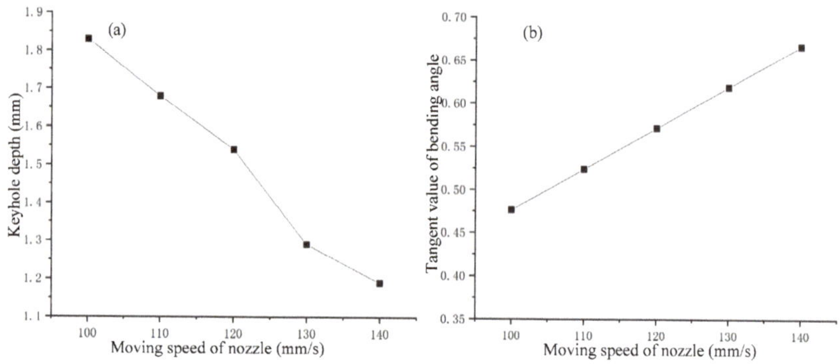

Figure 8. Effect of nozzle moving speed on keyhole: (**a**) Effect on keyhole depth; (**b**) Effect on keyhole bending.

3.2.2. Effect of Distance between Nozzle Outlet and Liquid Level on Keyhole

The nozzle was stationary, the gas flow rate was 81.5 mL/min, the distances between nozzle outlet and liquid level were 3 mm, 4 mm, 5 mm, 6 mm and 7 mm, respectively. The variation of keyhole depth with time is shown in Figure 9. It can be seen that when the distance reduced, the keyhole drilling rate in the fast-drilling stage also increased, but when the quasi-steady state stage was reached, the fluctuation of the keyhole would also become more violent. The change of the distance between nozzle outlet and liquid level was actually the change of the gas velocity on the liquid surface and the change in the gas–liquid interaction area. The effect of distance variation in the analogy welding was similar to the effect of defocusing variation in deep penetration laser welding.

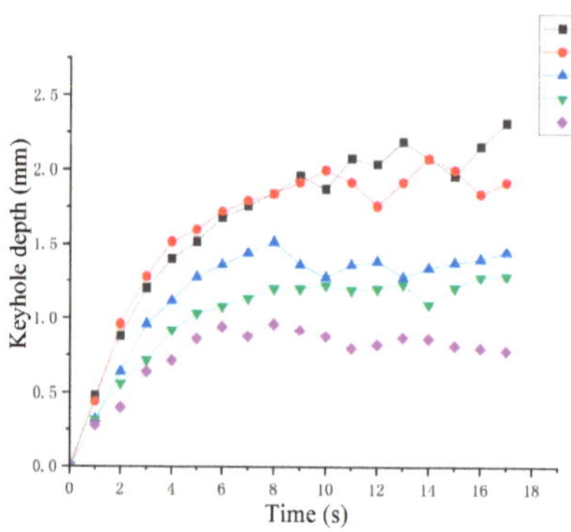

Figure 9. Effect of distance between nozzle outlet and liquid level on keyhole depth.

3.2.3. Effect of Material Properties on Keyhole

As shown in Figure 3, the surface tension coefficient of NaCl solution was linearly correlated with solute concentration. The surface tension coefficient is an important parameter in Formula (4). A series of NaCl solutions with different surface tension coefficients were prepared, which were impacted by certain gas flows, respectively. The obtained keyhole depth of each solution at the quasi-steady state stage with different surface tension coefficients is shown in Figure 10. Obviously, the keyhole depth was negatively correlated with the surface tension coefficient.

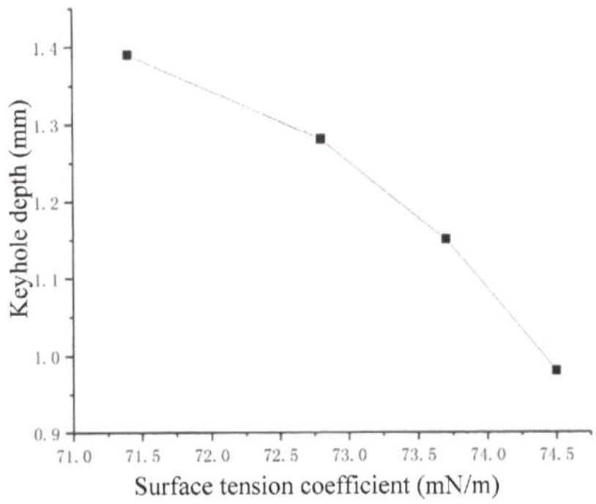

Figure 10. Keyhole depth corresponding to different surface tension coefficients.

According to Formula (2), the effect of surface tension was related to the radius of curvature, so the smaller the radius of curvature, the greater the influence of surface tension. When NaCl solutions with different mass fractions were used in the experiment, the density changed little, and it was considered to be a constant. The surface tension had the ability to close the keyhole, by calculation, and it could be seen that the additional pressure caused by the surface tension was much greater than the hydrostatic pressure at the bottom of the keyhole, so it could be considered that the surface tension of the material was the main factor determining the maximum depth of the keyhole.

Considering that the keyhole in laser welding is surrounded by molten metal, liquid metal material was selected as the experimental material for the analogy welding. For experimental safety, the alloy of Ga–In–Sn was selected as unselected mercury. Figure 11 shows the keyhole shape in the Ga–In–Sn alloy and the keyhole is outlined with red line. The keyhole in Figure 11a was obtained when the gas flow was 300 mL/min and the distance between nozzle outlet and liquid level was 3 mm. It can be seen that when the selected material physical properties were relatively close to those of the real molten pool material, the keyhole shape obtained by the analogy welding was similar to an inverted cone with a wide upper part and a narrow lower part.

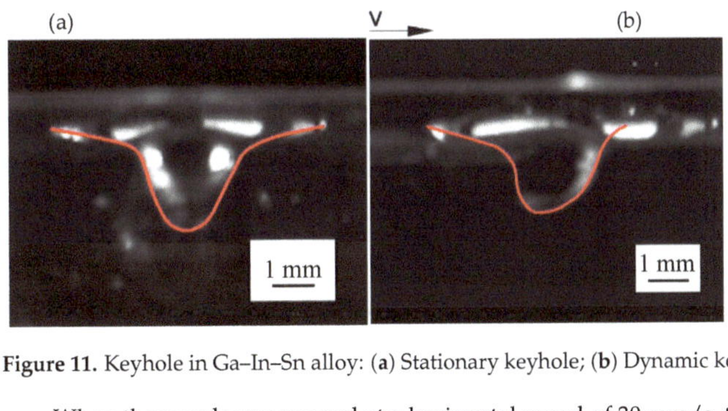

Figure 11. Keyhole in Ga–In–Sn alloy: (**a**) Stationary keyhole; (**b**) Dynamic keyhole.

When the nozzle was moved at a horizontal speed of 30 mm/s, the keyhole obtained is shown in Figure 11b. The front wall of the keyhole bent in the opposite direction of the moving speed, and the bending degree increased with increase in the moving speed.

When the distance between nozzle outlet and liquid level changed, the drawn curve of the maximum depth of the keyhole is shown in Figure 12. The effect of distance variation on keyhole depth in liquid metal was similar to the effect of defocusing variation in laser processing.

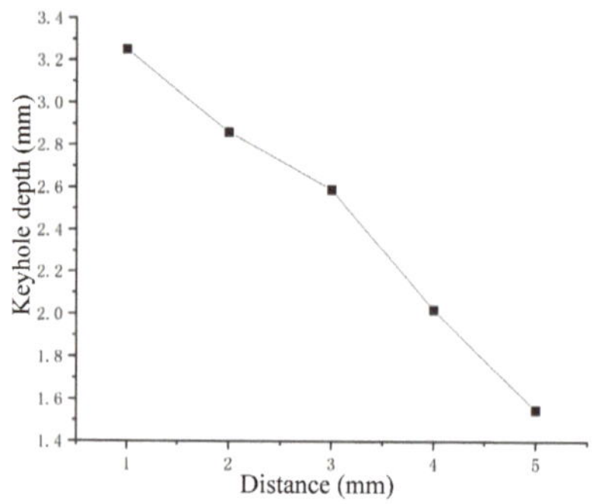

Figure 12. Keyhole depth in Ga–In–An alloy corresponding to different distance of nozzle.

The properties of liquid metal Ga–In–Sn alloy were similar to those of molten metal in welding. It may be of reference significance to explore the keyhole formed by gas blowing in analogy welding to study the keyhole in laser deep penetration welding.

3.3. *Keyhole in Real Deep Penetration Laser Welding with Sandwich Method*

The keyhole behavior in real deep penetration laser welding with the sandwich method was captured by the high-speed camera. The adopted camera frame rate was 10,000 fps, and several attenuation elements were used. The set laser power was 2500 W and the defocusing was 0 mm. Different welding speeds were set, and the obtained keyholes are shown in Figure 13. It can be seen that all keyholes at different welding speeds were bent in the opposite direction of the welding speed. A straight line passes through the top point of the keyhole front wall and is tangent to the keyhole front wall. The angle θ between that straight line and the vertical line is defined as the bending angle of the keyhole. The relationship between welding speed and keyhole depth and keyhole bending angle is shown in Figure 14. Comparing Figure 8 with Figure 14, it can be seen that the variation rules of keyhole depth and keyhole bending were basically the same in the analogy welding and in real welding. During a welding process, the variation of keyhole depth is shown in Figure 15. Comparing Figure 15 with Figure 5a, it can be seen that the variation trend of keyhole depth was similar. The stages of keyhole depth change in Figure 15 could still be divided into fast-drilling stage, slow-drilling stage and quasi-steady state stage.

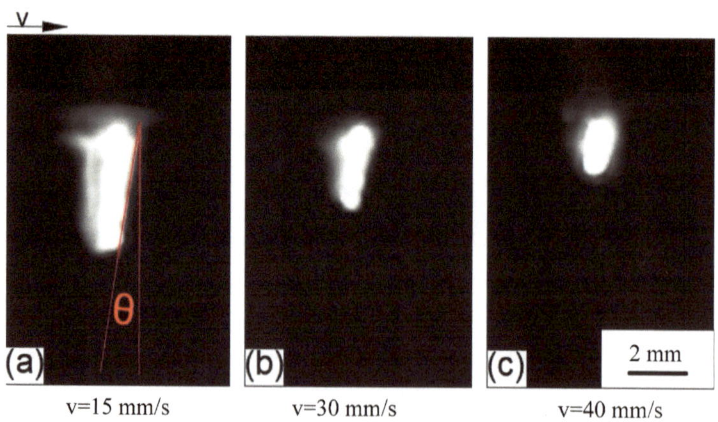

Figure 13. Keyhole shape at different speeds observed by the sandwich method: (**a**) Keyhole shape at welding speed of 15 mm/s; (**b**) Keyhole shape at welding speed of 30 mm/s; (**c**) Keyhole shape at welding speed of 40 mm/s.

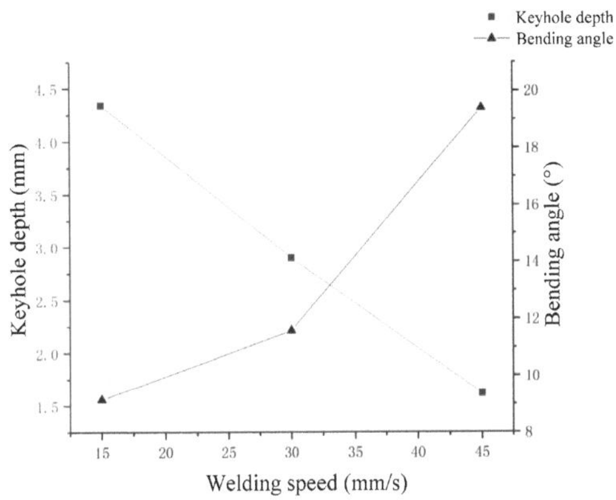

Figure 14. Effect of welding speed on keyhole depth and bending angle.

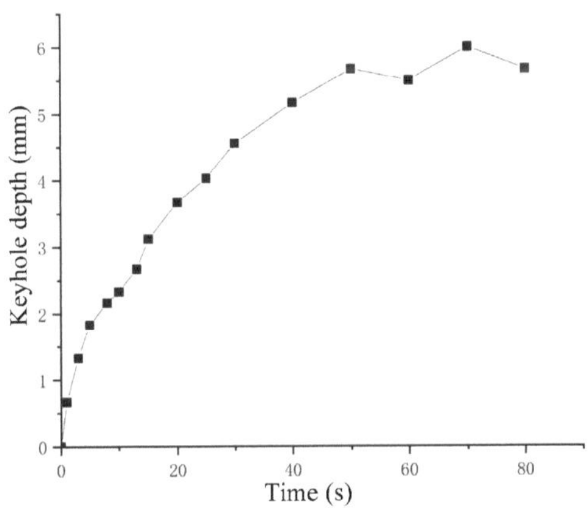

Figure 15. The variation of keyhole depth with time.

Figure 16a shows the keyhole obtained in real deep penetration laser welding when the defocusing was 5 mm, and Figure 16b shows the keyhole in analogy welding when the distance between nozzle outlet and liquid level was 5 mm. It can be seen that the two keyholes were very similar, both of them being slightly slender in shape. The front and rear walls in the keyhole middle part were smooth, while the bottoms of the two keyholes were curved and semicircular. In addition, the dynamic behavior of the two keyholes was similar. For example, the keyholes bent in the opposite direction of the welding speed, and the faster the speed was, the greater the bending degree was, and the smaller the keyhole depth was.

The main driving force of the keyhole generated by deep penetration laser welding is the vapor pressure generated by material evaporation of laser irradiated materials, while the force generating the keyhole in the analogy welding is mainly the dynamic pressure generated by the gas jet. Comparing Formula (3) with Formula (5), it can be seen that the composition of comprehensive pressure in the two kinds of keyholes is similar in form despite that the physical meanings being different. Mechanical properties influence the characteristics of keyholes. There are many similarities in the shape and dynamic behavior of keyholes obtained by analogy welding and real deep penetration laser welding, and the influence of process parameters on them is also similar. The study of keyhole formation and the maintenance mechanism obtained by analogy welding is helpful to understand the relevant mechanism in the real deep penetration laser welding process.

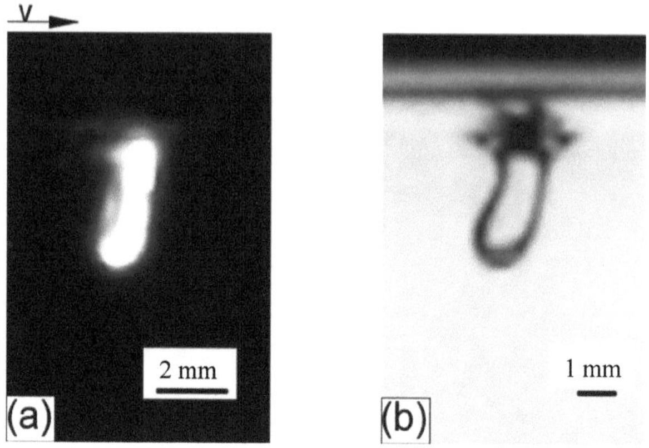

Figure 16. Obtained keyhole: (**a**) Keyhole in real deep penetration laser welding; (**b**) Keyhole in analogy welding.

4. Conclusions

In this paper, the keyholes in analogy welding and deep penetration laser welding were directly observed, and the effects of experimental parameters and material properties on the behavior of the keyholes were studied. The keyhole formed by gas blowing was compared with the keyhole generated in deep penetration laser welding. The following conclusions could be made:

(1) In the analogy welding and real deep penetration laser welding with the sandwich method, the pressure balance on the keyhole is similar in form. The gas dynamic pressure acting on the liquid in analogy welding and the vapor pressure caused by the evaporation of the laser irradiated material in deep penetration laser welding both promote the opening of the keyhole, while the hydrostatic pressure and surface tension promote the closing of the keyhole.
(2) When the process parameters in the analogy welding and real laser welding are used to study the influence on the keyhole, the behavior of the keyhole in the analogy welding is similar to that obtained in the real deep penetration laser welding.
(3) Studying the keyhole formation and maintenance principle in analogy welding may be helpful to deeply understand the keyhole formation and maintenance mechanism in real deep penetration laser welding.

Author Contributions: Conceptualization, X.J. and Z.H.; methodology, X.J.; software, H.C.; validation, Z.H., H.C. and Z.L.; formal analysis, Z.H.; investigation, H.C.; resources, X.J.; data curation, X.J.; writing—original draft preparation, Z.H.; writing—review and editing, X.J.; visualization, H.C.; supervision, Z.L.; project administration, X.J.; funding acquisition, X.J. All authors have read and agreed to the published version of the manuscript.

Funding: This research was funded by the National Natural Science Foundation of China, grant number 51875189.

Institutional Review Board Statement: Not applicable.

Informed Consent Statement: Not applicable.

Data Availability Statement: The data that support the findings of this study are available upon reasonable request from the authors.

Conflicts of Interest: The authors declare no conflict of interest.

Nomenclature

T	Surface tension
σ	Surface tension coefficient
L_b	Total length of boundary
ΔP	Additional pressure
R_1	Radius of curvature
R_2	Radius of curvature
P	Comprehensive pressure on gas-liquid interface
P_a	Atmospheric pressure
P_d	Dynamic pressure generated by gas jet
P_S	Hydrostatic pressure
p_0	Gas pressure in the nozzle
T_0	Gas temperature in the nozzle
ρ_0	Gas density in the nozzle
p_1	Gas pressure at the nozzle outlet
T_1	Gas temperature at the nozzle outlet
ρ_1	Gas density at the nozzle outlet
T_g	Temperature of jetted gas
ρ_g	Density of jetted gas
H	Distance between nozzle outlet and liquid level
h_0	Keyhole depth
R	Radius of curvature at the depth h on the keyhole wall
n	Normal of a point on the keyhole wall D Outlet diameter of the nozzle
v_g	Velocity of the gas in the jet prior to the collision with the surface of the liquid
ρ_L	Density of liquid
σ_L	Surface tension coefficient of liquid
R_{kh}	Radius of curvature of the liquid surface at the keyhole depth h_0
P_w	Comprehensive pressure in keyhole in laser welding
P_v	Vapor pressure of the material
v	Movement speed
V	Gas velocity at the nozzle outlet
q	Gas flow rate
θ	Angle
v_h	Moving speed of keyhole
v_d	Drilling speed of keyhole
F_H	The resultant force directions of the surface tension in the transverse direction
F_L	The resultant force directions of the surface tension in longitudinal direction

References

1. Arata, Y.; Abe, N.; Oda, T. Fundamental Phenomena in High Power CO_2 Laser Welding (Report I): Atmospheric Laser Welding (Welding Physics, Process&Instrument). *Trans. JWRI* **1985**, *14*, 5–11.
2. Dowden, J.; Postacioglu, N.; Davis, M.; Kapadia, P. A keyhole model in penetration welding with a laser. *J. Phys. D Appl. Phys.* **1987**, *20*, 36. [CrossRef]
3. Postacioglu, N.; Kapadia, P.; Davis, M.; Dowden, J. Upwelling in the liquid region surrounding the keyhole in penetration welding with a laser. *J. Phys. D Appl. Phys.* **1987**, *20*, 340. [CrossRef]
4. Dowden, J.; Kapadia, P.; Clucas, A.; Ducharme, R.; Steen, W.M. On the relation between fluid dynamic pressure and the formation of pores in laser keyhole welding. *J. Laser Appl.* **1996**, *8*, 183–190. [CrossRef]
5. Postacioglu, N.; Kapadia, P.; Dowden, J.M. The thermal stress generated by a moving elliptical weldpool in the welding of thin metal sheets. *J. Phys. D Appl. Phys.* **1997**, *30*, 2304. [CrossRef]
6. Dowden, J. Interaction of the keyhole and weld pool in laser keyhole welding. *J. Laser Appl.* **2002**, *14*, 204–209. [CrossRef]
7. Semak, V.V.; Hopkins, J.A.; McCay, M.H.; McCay, T.D. Melt pool dynamics during laser welding. *J. Phys. D Appl. Phys.* **1995**, *28*, 2443. [CrossRef]
8. Semak, V.; Matsunawa, A. The role of recoil pressure in energy balance during laser materials processing. *J. Phys. D Appl. Phys.* **1997**, *30*, 2541. [CrossRef]
9. Kim, J.; Oh, S.; Ki, H. A study of keyhole geometry in laser welding of zinc-coated and uncoated steels using a coaxial observation method. *J. Mater. Process. Technol.* **2015**, *225*, 451–462. [CrossRef]
10. Lankalapalli, K.N.; Tu, J.F.; Gartner, M. A model for estimating penetration depth of laser welding processes. *J. Phys. D Appl. Phys.* **1996**, *29*, 1831. [CrossRef]

11. Solana, P.; Negro, G. A study of the effect of multiple reflections on the shape of the keyhole in the laser processing of materials. *J. Phys. D Appl. Phys.* **1997**, *30*, 3216. [CrossRef]
12. Pecharapa, W.; Kar, A. Effects of phase changes on weld pool shape in laser welding. *J. Phys. D Appl. Phys.* **1997**, *30*, 3322. [CrossRef]
13. Strömbeck, P.; Kar, A. Self-focusing and beam attenuation in laser materials processing. *J. Phys. D Appl. Phys.* **1998**, *31*, 1438. [CrossRef]
14. Fabbro, R.; Chouf, K. Keyhole modeling during laser welding. *J. Appl. Phys.* **2000**, *87*, 4075–4083. [CrossRef]
15. Fabbro, R.; Slimani, S.; Coste, F.; Briand, F. Analysis of the Various Melt Pool Hydrodynamic Regimes Observed during cw nd-yag Deep Penetration Laser Welding. In Proceedings of the 26th International Congress on Applications of Lasers and Electro-Optics, Orlando, FL, USA, 29 October–1 November 2007.
16. Fabbro, R.; Slimani, S.; Coste, F.; Briand, F. Characteristic Melt Pool Hydrodynamic Behaviors for CW Nd-YAG Deep Penetration Laser Welding. In Proceedings of the 3rd Pacific International Conference on Applications of Lasers and Optics, Beijing, China, 16–18 April 2008.
17. Fabbro, R. Melt pool and keyhole behaviour analysis for deep penetration laser welding. *J. Phys. D Appl. Phys.* **2010**, *43*, 445501. [CrossRef]
18. Fabbro, R. Scaling laws for the laser welding process in keyhole mode. *J. Mater. Process. Technol.* **2019**, *264*, 346–351. [CrossRef]
19. Jin, X.; Li, L.; Zhang, Y. A heat transfer model for deep penetration laser welding based on an actual keyhole. *Int. J. Heat Mass Transf.* **2003**, *46*, 15–22. [CrossRef]
20. Li, L.; Jin, X.; Zhang, Y. A Theoretical and Experimental Study on Keyhole Effects in Laser Deep-Penetration Welding. In Proceedings of the Lasers in Material Processing and Manufacturing, Shanghai, China, 14–18 October 2002.
21. Jin, X.Z.; Li, L.J. A conduction model for deep penetration laser welding based on an actual keyhole. *Opt. Laser Technol.* **2003**, *35*, 5–12. [CrossRef]
22. Jin, X.; Berger, P.; Graf, T. Multiple reflections and Fresnel absorption in an actual 3D keyhole during deep penetration laser welding. *J. Phys. D Appl. Phys.* **2006**, *39*, 4703. [CrossRef]
23. Jin, X.; Li, L. An experimental study on the keyhole shapes in laser deep penetration welding. *Opt. Laser Eng.* **2004**, *41*, 779–790. [CrossRef]
24. Jin, X. A three-dimensional model of multiple reflections for high-speed deep penetration laser welding based on an actual keyhole. *Opt. Lasers Eng.* **2008**, *46*, 83–93. [CrossRef]
25. Zhang, Y.; Chen, G.; Wei, H.; Zhang, J. A novel "sandwich" method for observation of the keyhole in deep penetration laser welding. *Opt. Lasers Eng.* **2008**, *46*, 133–139. [CrossRef]
26. Cheng, Y.; Jin, X.; Li, S.; Zeng, L. Fresnel absorption and inverse bremsstrahlung absorption in an actual 3D keyhole during deep penetration CO_2 laser welding of aluminum 6016. *Opt. Laser Technol.* **2012**, *44*, 1426–1436. [CrossRef]
27. Jin, X.; Zeng, L.; Cheng, Y. Direct observation of keyhole plasma characteristics in deep penetration laser welding of aluminum alloy 6016. *J. Phys. D Appl. Phys.* **2012**, *45*, 245205. [CrossRef]
28. Li, S.; Chen, G.; Zhang, M.; Zhou, Y.; Zhang, Y. Dynamic keyhole profile during high-power deep-penetration laser welding. *J. Mater. Process. Technol.* **2014**, *214*, 565–570. [CrossRef]
29. Zhang, Y.; Lin, Q.; Yin, X.; Li, S.; Deng, J. Experimental research on the dynamic behaviors of the keyhole and molten pool in laser deep-penetration welding. *J. Phys. D Appl. Phys.* **2018**, *51*, 145602. [CrossRef]
30. Zhang, D.; Wang, M.; Shu, C.; Zhang, Y.; Wu, D.; Ye, Y. Dynamic keyhole behavior and keyhole instability in high power fiber laser welding of stainless steel. *Opt. Laser Technol.* **2019**, *114*, 1–9. [CrossRef]
31. Wang, H.; Shi, Y.; Gong, S. Numerical simulation of laser keyhole welding processes based on control volume methods. *J. Phys. D Appl. Phys.* **2006**, *39*, 4722. [CrossRef]
32. Cho, J.H.; Farson, D.F.; Milewski, J.O.; Hollis, K.J. Weld pool flows during initial stages of keyhole formation in laser welding. *J. Phys. D Appl. Phys.* **2009**, *42*, 175502. [CrossRef]
33. Huang, B.; Chen, X.; Pang, S.; Hu, R. A three-dimensional model of coupling dynamics of keyhole and weld pool during electron beam welding. *Int. J. Heat Mass Transf.* **2017**, *115*, 159–173. [CrossRef]
34. Bedenko, D.V.; Kovalev, O.B.; Krivtsun, I.V. Simulation of plasma dynamics in a keyhole during laser welding of metal with deep penetration. *J. Phys. D Appl. Phys.* **2010**, *43*, 105501. [CrossRef]
35. Pang, S.; Chen, X.; Shao, X.; Gong, S.; Xiao, J. Dynamics of vapor plume in transient keyhole during laser welding of stainless steel: Local evaporation, plume swing and gas entrapment into porosity. *Opt. Lasers Eng.* **2016**, *82*, 28–40. [CrossRef]
36. Li, L.; Peng, G.; Wang, J.; Gong, J.; Meng, S. Numerical and experimental study on keyhole and melt flow dynamics during laser welding of aluminium alloys under subatmospheric pressures. *Int. J. Heat Mass Transf.* **2019**, *133*, 812–826. [CrossRef]
37. Cunningham, R.; Zhao, C.; Parab, N.; Kantzos, C.; Pauza, J.; Fezzaa, K.; Sun, T.; Rollett, A.D. Keyhole threshold and morphology in laser melting revealed by ultrahigh-speed x-ray imaging. *Science* **2019**, *363*, 849–852. [CrossRef] [PubMed]
38. Mayi, Y.A.; Dal, M.; Peyre, P.; Bellet, M.; Metton, C.; Moriconi, C.; Fabbro, R. Transient dynamics and stability of keyhole at threshold in laser powder bed fusion regime investigated by finite element modeling. *J. Laser Appl.* **2021**, *33*, 012024. [CrossRef]
39. Zou, J.; Ha, N.; Xiao, R.; Wu, Q.; Zhang, Q. Interaction between the laser beam and keyhole wall during high power fiber laser keyhole welding. *Opt. Express* **2017**, *25*, 17650–17656. [CrossRef]

40. Li, Z.; Yu, G.; He, X.; Li, S.; Tian, C.; Dong, B. Analysis of surface tension driven flow and solidification behavior in laser linear welding of stainless steel. *Opt. Laser Technol.* **2020**, *123*, 105914. [CrossRef]
41. Seidgazov, R.D. Analysis of The Main Hydrodynamic Mechanisms in Laser Induced Keyhole Welding. In Proceedings of the IEEE 8th International Conference on Advanced Optoelectronics and Lasers, Sozopol, Bulgaria, 6–8 September 2019.
42. Wu, D.; Hua, X.; Huang, L.; Li, F.; Cai, Y. Observation of the keyhole behavior, spatter, and keyhole-induced bubble formation in laser welding of a steel/glass sandwich. *Weld. World* **2019**, *63*, 815–823. [CrossRef]
43. Huang, Y.; Shen, C.; Ji, X.; Li, F.; Zhang, Y.; Hua, X. Correlation between gas-dynamic behaviour of a vapour plume and oscillation of keyhole size during laser welding of 5083 Al-alloy. *J. Mater. Process. Technol.* **2020**, *283*, 116721. [CrossRef]
44. Hao, Z.; Chen, H.; Jin, X.; Yu, X.; Liu, Z. Direct observation of keyhole with an innovative layer-by-layer imaging method during deep penetration laser welding. *Opt. Laser Technol.* **2022**, *152*, 108142. [CrossRef]
45. Olivares, O.; Elias, A.; Sanchez, R.; Díaz-Cruz, M.; Morales, R.D. Physical and mathematical models of gas-liquid fluid dynamics in LD converters. *Steel Res.* **2002**, *73*, 44–51. [CrossRef]
46. Vafaei, S.; Wen, D. Modification of the Young–Laplace equation and prediction of bubble interface in the presence of nanoparticles. *Adv. Colloid Interface Sci.* **2015**, *225*, 1–15. [CrossRef] [PubMed]
47. Banks, R.B.; Chandrasekhara, D.V. Experimental investigation of the penetration of a high-velocity gas jet through a liquid surface. *J. Fluid Mech.* **1963**, *15*, 13–34. [CrossRef]
48. Banks, R.B.; Bhavamai, A. Experimental study of the impingement of a liquid jet on the surface of a heavier liquid. *J. Fluid Mech.* **1965**, *23*, 229–240. [CrossRef]
49. Mordasov, M.M.; Savenkov, A.P.; Chechetov, K.E. Method for analyzing the gas jet impinging on a liquid surface. *Technol. Phys.* **2016**, *61*, 659–668. [CrossRef]
50. Szymanski, Z.; Hoffman, J.; Kurzyna, J.; Kolasa, A.; Jakubowski, J. Oscillations of keyhole pressure and plasma radiation during cw CO_2 laser welding. *Laser Technol. VI Appl.* **2000**, *4238*, 186–198.
51. Vargaftik, N.B.; Volkov, B.N.; Voljak, L.D. International tables of the surface tension of water. *J. Phys. Chem. Ref. Data* **1983**, *12*, 817–820. [CrossRef]
52. Hwang, H.Y.; Irons, G.A. Mathematical modeling of impinging gas jets on liquid surfaces. *Metall. Mater. Trans. B* **2011**, *42*, 575–591. [CrossRef]
53. Kestin, J.; Khalifa, H.E.; Correia, R.J. Tables of the dynamic and kinematic viscosity of aqueous NaCl solutions in the temperature range 20–150 C and the pressure range 0.1–35 MPa. *J. Phys. Chem. Ref. Data* **1981**, *10*, 71–88. [CrossRef]
54. dos Santos, A.P.; Diehl, A.; Levin, Y. Surface tensions, surface potentials, and the Hofmeister series of electrolyte solutions. *Langmuir* **2010**, *26*, 10778–10783. [CrossRef]
55. Plevachuk, Y.; Sklyarchuk, V.; Eckert, S.; Gerbeth, G.; Novakovic, R. Thermophysical properties of the liquid Ga–In–Sn eutectic alloy. *J. Chem. Eng. Data* **2014**, *59*, 757–763. [CrossRef]
56. Ek, M.; Sichen, D. Study of penetration depth and droplet behavior in the case of a gas jet impinging on the surface of molten metal using liquid Ga–In–Sn. *Steel Res. Int.* **2012**, *83*, 678–685. [CrossRef]
57. Liu, Q.; Chen, W.; Hu, L.; Xie, H.; Fu, X. Experimental investigation of cavity stability for a gas-jet penetrating into a liquid sheet. *Phys. Fluids* **2015**, *27*, 082106. [CrossRef]

materials

Article

Analytical Simulation of the Microbubble Collapsing in a Welding Fusion Pool

Ahmed Teyeb [1,*], **Mohamad Salimi** [1], **Evelyne El Masri** [1], **Wamadeva Balachandran** [2] **and Tat-Hean Gan** [1,3,*]

[1] Brunel Innovation Centre, Brunel University London, Uxbridge UB8 3PH, UK
[2] Department of Electronic and Electrical Engineering, Brunel University London, Uxbridge UB8 3PH, UK
[3] TWI Ltd., Granta Park, Great Abignton, Cambridge CB 21 6AL, UK
* Correspondence: ahmed.teyeb@brunel.ac.uk (A.T.); tat-hean.gan@brunel.ac.uk (T.-H.G.)

Abstract: This paper explains the use of remote ultrasound vibration at the optimum position and frequencies to vibrate plates under welding, with the aim of initiating cavitation in the molten pool area. It has been shown in the literature that ultrasound cavitation changes microstructure morphology and refines the grain of the weld. In practice, the plates are excited through narrow-band high-power ultrasound transducers (HPUTs). Therefore, a theoretical investigation is carried out to identify the plate-mode shapes due to the ultrasound vibration aligned with the frequency bandwidth of HPUTs available in the marketplace. The effect of exciting the plate at different locations and frequencies is studied to find the optimum position and frequencies to achieve the maximum pressure at the area of the fusion zone. It was shown that applying the excitation from the side of the plate produces an order of 10^3 higher vibration displacement amplitude, compared with excitation from the corner. The forced vibration of cavitation and bursting time are studied to identify vibration amplitude and the time required to generate and implode cavities, hence specifying the vibration-assisted welding time. Thus, the proposed computational platform enables efficient multiparametric analysis of cavitation, initiated by remote ultrasound excitation, in the molten pool under welding.

Keywords: vibration assisted welding; power ultrasonic; cavitation

Citation: Teyeb, A.; Salimi, M.; El Masri, E.; Balachandran, W.; Gan, T.-H. Analytical Simulation of the Microbubble Collapsing in a Welding Fusion Pool. *Materials* **2023**, *16*, 410. https://doi.org/10.3390/ma16010410

Academic Editors: Swarup Bag and Christ Prakash Paul

Received: 1 December 2022
Revised: 27 December 2022
Accepted: 28 December 2022
Published: 1 January 2023

1. Introduction

Most manufacturing processes, such as laser welding, additive manufacturing, and casting, are essentially based on the phenomenon of total fusion of material, followed by rapid solidification. The material's structure in the final phase and its mechanical and electrical properties strongly depend on the solidification phase [1,2]. This is especially evident when the material consists of an alloy of metals or even a combination of two dissimilar materials, such as in the case of lap joining (by laser welding) of specific connectors to battery cells, for the manufacturing of battery packs for electric vehicles [3]. The difference in melting points between different types of metals leads to the coexistence of liquid and solid phases and subsequently to the agglomeration of particles of similar nature, which ideally should be dispersed homogeneously throughout the liquid phase during solidification [4,5].

The presence of voids and gas, and the grain size variability, are phenomena frequently observed that cause the final material's properties to deteriorate, leading to problems such as hot cracking and, subsequently, fragile connection [6]. Ultrasonic processing of molten materials is the primary technique that helps to improve the microstructure by grain refinement. The vibration can be transmitted to the molten pool in two ways: vibration of the workpiece [7–9] or tool vibration [10–12]. Four mechanisms, cavitation, acoustic flow, mechanical effect, and thermal effect, are produced in the fusion zone during ultrasonic vibration. One of the remarkable applications of high-power ultrasound on molten metal is the creation of cavitation in the melt. Acoustic cavitation is a powerful phenomenon promoting melt microheterogeneity and the main processes of degassing and fine filtration in light alloys. The acoustic cavitation in the molten pool contributes to the fragmentation and shaping of intermetallic compounds, which is the leading cause for the degradation of the properties [13]. Four mechanisms are involved in acoustic cavitation: acoustic streaming, microstreaming, microjets, and microstreams.

The associated phenomenon is described as follows. The cavitation bubble expands and contracts in the molten pool until it is exposed to the high-pressure region from the ultrasound wave and then implodes. The implosion of a bubble generates a high-pressure acoustic wave called a shock wave and a local hot spot at a very high temperature. The other mechanisms, such as acoustic streaming, microstreaming, microstreamers, and microjets, are followed by the implosion of the cavitation [14].

The effect of vibration and waves on the creation of a cavitation bubble has been presented in several studies [15,16]. The emission of the shock wave due to the collapse of a cavitation bubble attached to a rigid wall was investigated by Brujan et al. [17]. Their results indicated that a significant portion of the shock wave is dissipated within 100 μm from the bubble wall. Ultrasound cavitation can burst under the compression of the ultrasound wave sooner than they fill with dissolved gas in the melt. Typically, the bubble threshold for collapsing increases as the frequency of ultrasound increases; 1.2 atm at 20 kHz, 1.8 atm at 140 kHz, 3 atm at 1 MHz, and 5.8 atm at 5 MHz [15,16].

Several studies were carried out to characterise the effect of ultrasonic waves on inert tungsten gas (TIG) welding [18–21]. It was shown by Sun et al. [18] that there is a 300% increase in the penetration rate compared with conventional TIG welding. The increase in the penetration results from increased arc push force via ultrasound, leading to the oscillation in the plasma arc. Ultrasonic-assisted laser welding has been studied by several scholars, for example, [22–26]. Finite element modelling and a laser Doppler vibrometer were used to estimate the ultrasonic energy propagation on the workpiece by Tarasov et al. [26]. An improvement in the microstructure, microhardness, and tensile strength of the 321 stainless steel weld was reported when ultrasonic vibration is transmitted into the welding zone. Lei et al. [25] reported that the weld porosity decreased from 4.3% to 0.9% with a reduction in its average size. Teyeb et al. [27] showed experimentally that weld strength increased by 26% when the laser welding was assisted by ultrasonic vibration. Deeper penetration is reported by Woizeschke et al. [28] and Radel [29] when laser welding is assisted by ultrasonic vibration, which is due to the heating effect of ultrasonic vibration.

The effect of using different ultrasound power on the porosity of a welding joint is experimentally investigated by Yin et al. [30]. The results showed that appropriate ultrasonic energy significantly increases the grain state of the weld. The tensile strength of a welded joint increased by approximately 12% compared with conventional welding. Increasing the ultrasonic power beyond the appropriate level can degrade the tensile strength. As the ultrasound melt treatment before solidification notably affects refining and solidification in light alloys, the welding can be called microcasting.

In this paper, the theoretical vibration of the plate coupled with cavitation vibration in the molten area is presented. As the HPUTs available in the marketplace can excite the plate in a region of narrow band frequency, a simulation study was carried out to understand the wave propagation and modes in the structures at the desired frequencies. An objective of this study is to find the optimum position to remotely vibrate the centre of the plate with the maximum amplitude and hence generate ultrasound capitations in the molten pool area. The cavitation phenomenon that occurs during the solidifying phase is the origin of the microstructure improvements in the joints. Therefore, further theoretical investigations were conducted to find the cavitation vibration at the molten pool area, initiated and vibrated by the remote excitation of the plate. Such a theoretical investigation can help to find the position, frequency, and time for vibration-assisted welding.

2. Theoretical Modelling

This section provides theoretical explanations of the plate's vibration under ultrasonic-guided wave propagation and its interactions with the ultrasound cavitation in the fusion zone. As shown in Figure 1, part of the ultrasound-guided wave travelling in the plate leaks or refracts into the liquid. When the sound pressure of the UGW (ultrasonic-guided waves) exceeds the cavitation threshold, several microbubbles are generated in the fluid close to the solid–fluid boundary. As the ultrasound excites the plate, the microbubbles are subjected to positive and negative pressure. They expand and contract under negative and positive pressure, respectively [31,32].

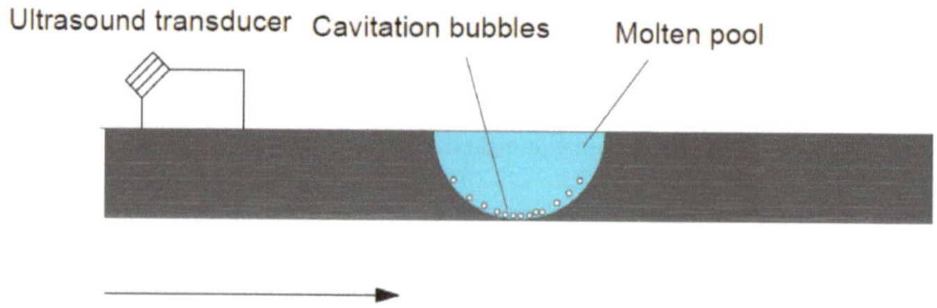

Figure 1. Leakage of the guided wave into the fluid. The water boils due to low local pressure, which generates cavitation.

2.1. Ultrasonic Plate Vibration

Modelling of plate vibration has been studied extensively since 1787 [33]. Most of the relevant background theory on the free and forced plate vibration can be found in the reference work by Leisaa [34]. Owing to the presence of the higher order matrices, the virtual approach is suggested by Sung et al. [35–37] in a series of studies, making it computationally less expensive to estimate the response of the plate. In this study, the theoretical modelling from Vlasov [38] is used to obtain the mode shapes of the plate at the desired frequency region. Based on the principle of virtual work, the steady-state transverse displacement, $\xi(x,y)$ of a full clamped plate subjected to harmonic point excitation at $\xi(x',y')$ is:

$$\xi(x,y) = F_0 \sum_{m=1}^{\infty} \sum_{n=1}^{\infty} \frac{\Psi_{mn}(x,y)\Psi_{mn}(x',y')}{B(I_1 I_2 + 2I_3 I_4 + I_5 I_2 - \rho_s \omega^2 I_2 I_6)} \tag{1}$$

where F_0 is the force amplitude, B is the bending stiffness $B = \mathrm{Eh}^3/12(1-\mathrm{v}^2)$, E is Young's modulus, v is the Poisson ratio, ρ_s is the plate density and ω is the angular frequency.

The eigenfunctions can express the shape function associated with the plate:

$$\Psi_{mn}(x,y) = \vartheta_m(x)\zeta_n(y). \tag{2}$$

The ϑ_m and ζ_n parameters can be expressed by Bessel and Hankel functions, respectively:

$$\vartheta_m(x) = J\left(\frac{B_{mx}}{a}\right) - \frac{J(\beta_m)}{H(\beta_m)} H \frac{\beta_m x}{a,}, \zeta_n(y) = J\left(\frac{\beta_n y}{b}\right) - \frac{J(\beta_m)}{H(\beta_m)} H\left(\frac{\beta_n y}{b}\right), \tag{3}$$

where: $J(s) = \cosh(s) - \cos(s)$ and $H(s) = \sinh(s) - \sin(s)$, and β_n is the n-th root of $\cosh(\beta)\cos(\beta) = 0$. The I parameter in the denominator of Equation (1) can be expressed by:

$$I_1 = \int_0^a \vartheta_m''' \vartheta_m dx,\ I_2 = \int_0^b \zeta_n^2\, dy,\ I_3 = \int_0^a \vartheta_m'' \vartheta_m dx,\ I_4 = \int_0^a \zeta_n'' \zeta_n dx,\ I_5 = \int_0^b \zeta_n''' \zeta_n dy\ I_6 = \int_0^a \vartheta_m^2 dx. \tag{4}$$

The natural frequency of the plate can be estimated using the eigenfunctions:

$$\omega_{mn} = \sqrt{\frac{B}{\rho_s}}\sqrt{\frac{I_1 I_2 + 2I_3 I_4 + I_5 I_6}{I_2 I_6}}. \quad (5)$$

A high-power ultrasound transducer (HPUT) can generate ultrasonic compressional vibration at a narrow band frequency. It is advantageous to select the transducer to align with the resonance frequency of the plate that can generate the high-pressure acoustic wave in the fusion zone of the weld.

Figure 2 shows a 25 cm × 25 cm × 0.1 cm clamped aluminium alloy plate-mode shape with alloy density ρ_s = 2800 kg/m^3, Poisson ratio ν = 0.33 and Young modulus E = 72.5 GPa, subjected to unit force at the locations illustrated in Figure 3. Although in ultrasonic melting treatment different frequency bandwidth are employed to excite the plate, in this study the plate-mode shapes are plotted for the frequency 20 kHz, subjected to a unit force at the locations illustrated in Figure 3.

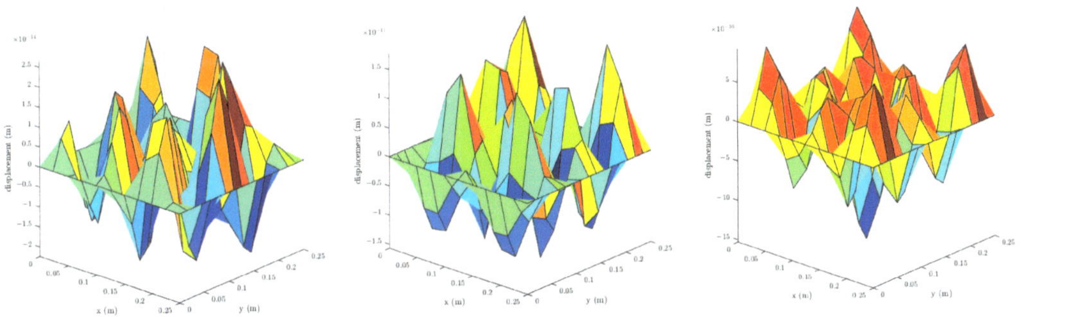

(a) (b) (c)

Figure 2. A clamped-plate vibration-mode shape subjected to unit force excitation at the (**a**) corner, (**b**) side and (**c**) centre of the plate at 20 kHz. The excitation locations are illustrated in Figure 3.

As illustrated in Figure 2, at the excitation point the displacement reaches its maximum value. Although using a vibration close to the tip of the laser head is preferable, it might not be practical to be very close to the laser head due to the heating produced in the welding zone area. Another approach is to use a fixed location for the ultrasound to vibrate the fusion zone area remotely. As illustrated in Figure 2b,c when the plate is subjected to excitation from the side, the displacement amplitude is an order of 10^3 higher compared to the corner excitation.

Figure 3. Clamped aluminium alloy plate subjected to ultrasonic excitation at three locations.

Ahmed et al. [27] used an ultrasound transducer at the side of a plate to transmit the ultrasound vibration into the molten pool. In this study, the effect of a single transducer on the plate displacement is plotted at the frequencies used by Teyeb et al. [27]: 19, 28 and 40 kHz. The excitation location is highlighted by a red arrow in Figure 3. For the excited plates, shown in Figure 4, the one subjected to 40 kHz has a mode shape where the middle of the plate has a minimum displacement amplitude compared with the other frequencies. The maximum displacement is associated with the plated subjected to 20 kHz excitation.

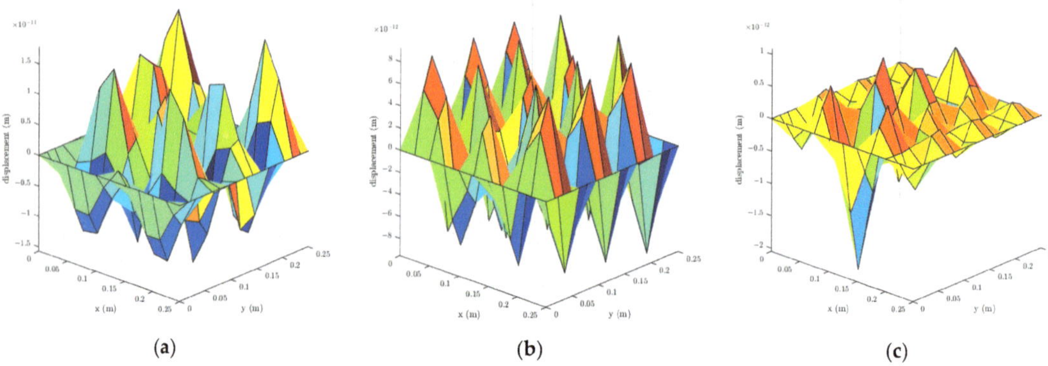

Figure 4. Excitation from the side using a single transducer at (**a**) 20, (**b**) 28 and (**c**) 40 kHz. The excitation location is highlighted by a red arrow in Figure 3.

2.2. *Coupling of the Plate Vibration and the Fusion Area Pressure*

To identify the pressure in the fusion zone, it is assumed that the vibration is applied to a semicylinder. The relation between internal pressure coefficient P_f and the circumferential displacement of a cylinder W can be expressed by [39]:

$$P_f = \frac{-2B_s}{1-\left(\alpha_s - \alpha_f\right)^2}\frac{W}{a}, \tag{6}$$

where B_s and a is the bulk modulus of the semifluid and the cylinder radius, respectively. The $\alpha_s = k_s a$ and $\alpha_f = k_f a$, where k_s and k_f are the axial wave number of the semifluid dominated wave and the semifluid wavenumber respectively. The pressure within the semifluid can be described by a Bessel function of order zero:

$$p = P_f J_0(k_s^r a), \tag{7}$$

where the k_s^r parameter is the radial wavenumber and is related to the semifluid wavenumber k_f by: $(k_s^r)^2 = k_f^2 - k_s^2$.

The pressure variation in the molten pool is plotted in Figure 5 using the displacement value associated with the centre of the plate shown in Figure 4, assumed 10^{-11} (m). As illustrated in Figure 5, the optimum frequencies for exciting the plate are approximately 20 and 28 kHz. Considering the results shown in Figures 4 and 5, it is recommended to use either a 20 or 28 kHz transducer.

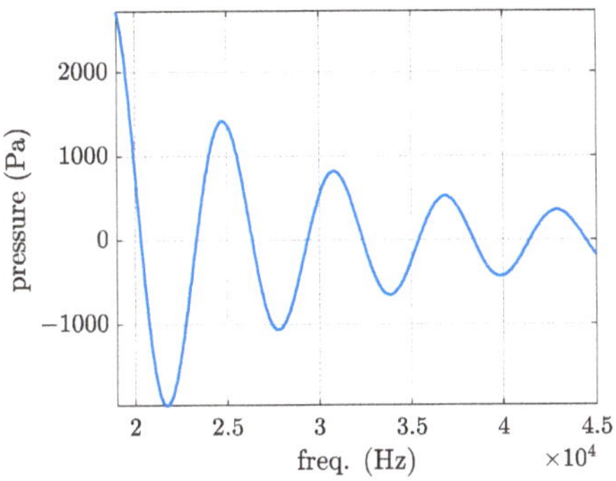

Figure 5. Pressure variation in the molten pool with respect to frequency.

2.3. Cavitation Dynamics

Cavitation is referred to the generation of cavities and the subsequent oscillation behaviour due to exceeding the binding force between the melt molecules [40]. The cavitation threshold in the melt is given by [41]:

$$P_B = P_0 - P_v + \frac{2}{3\sqrt{3}}\left[\frac{\left(\frac{2\sigma}{R_0}\right)^3}{P_0 - P_v + \frac{2\sigma}{R_0}}\right]^{\frac{1}{2}}, \tag{8}$$

where P_0 is the liquid static pressure, P_v is the vapour pressure in the cavitation and σ is the surface tension coefficient of the melt. The R_0 parameter is the initial radius of the cavitation.

The relationships between the initial radius of cavitation and the sound field frequency f_r can be expressed by [42]:

$$f_r = \frac{1}{2\pi R_0}\left[\frac{3\gamma}{\rho_f}\left(P_0 + \frac{2\sigma}{R_0}\right) - \frac{2\sigma}{\rho_f R_0}\right]^{\frac{1}{2}}, \tag{9}$$

where γ is the specific heat capacity of the fluid and ρ_f is the fluid density.

By assuming the physical parameters P_v = 2000 Pa, σ = 0.910 N/m, P_0 = 0.101 MPa, R_0 = 10 µm, and actual cavitation threshold P_B = 155,411 Pa, Equations (8) and (9) are plotted in Figure 6. As seen in Figure 6, increasing the frequency gives a smaller cavity but requires a higher pressure to initiate it. The required pressure to initiate cavitation is 100 times higher than the maximum value illustrated in Figure 5. Hence, using a mobile transcoder close to the area where the welding operation is happening is recommended to achieve the desired displacement level, as shown in Figure 2c.

The behaviour of cavities has been extensively studied and is covered in several literature references [40,42,43]. In this study, the oscillation of a vapour-gas cavity in an incompressible liquid is governed by the Nolting–Neppiras equation:

$$R\ddot{R} + \frac{3}{2}\dot{R}^2 + 4\mu\frac{\dot{R}}{R} + \frac{2\sigma}{R} = \left(p_0 - P_v + \frac{2\sigma}{R_0}\right)\left(\frac{R_0}{R}\right)^3 - p_0 + p_v + p_A\sin(\omega t) \tag{10}$$

where R, $\dot{R}$, $\ddot{R}$ are cavitation radius and its first and second derivative. The R_0, σ and μ terms are the initial radius of the cavity, the surface tension of the melt and the viscosity of the melt. The p_0, p_v and p_A parameters are ambient pressure, sound pressure and vapour pressure, respectively.

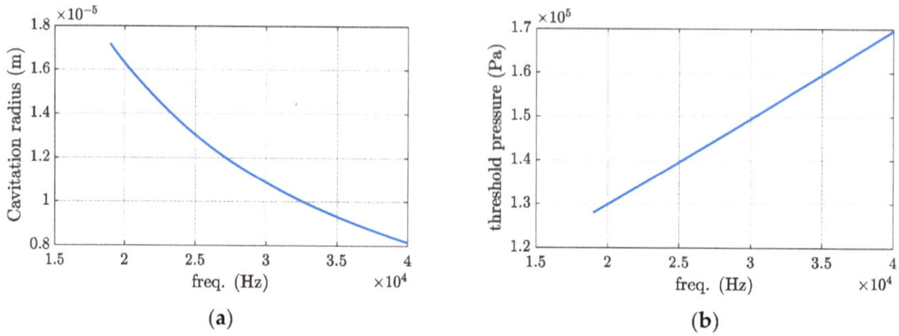

Figure 6. Changes to the cavitation (**a**) radius and (**b**) threshold pressure with respect to the excitation frequency.

The pressure relapsed by imploding the cavitation is proportional to the acoustic pressure of the melt. The higher the acoustic pressure of the melt, the higher the acoustic wave released by the cavitation. The critical radius can be estimated by:

$$Pg + P_v - P_0 = \frac{2\sigma}{R}, \tag{11}$$

where $Pg + P_v$ is the total pressure inside the bubble.

The pressure distribution around the ultrasound cavitation was studied by [43], and the results indicated that the pressure reaches its maximum value at about 1.58 R from the cavitation bubble. It is shown in [44] that the weld's penetration and width increased when pulsed ultrasound was employed.

Changes to the cavitation radius and the wall velocity are plotted in Figure 7, using Equation (10). As seen from the plots, the ultrasound cavitation oscillation happens up to approximately 100 milliseconds. The cavitation radius and the wall velocity can reach approximately 0.004 mm and 30 m/s, respectively. The pressure distribution around the cavitation bubble was studied in [43], and the results indicated that the pressure reaches its maximum value at about 1.58 R from the cavitation bubble. Based on the results shown in Figure 7a, the effect of the shock wave from each ultrasound cavitation would be approximately 6.3×10^{-3} mm in the molten pool area. Depending on the width and the depth of the fusion zone, different cavitation numbers should be initiated to have ultrasonic grain refining.

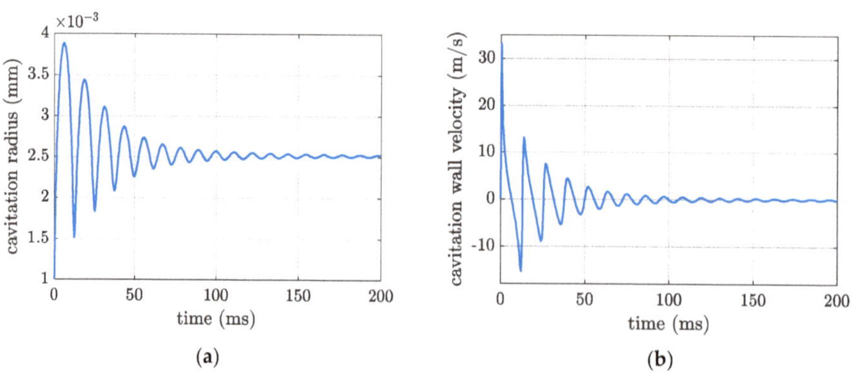

Figure 7. Cavitation (**a**) radius and (**b**) wall velocity subjected to 20 kHz vibration.

3. Conclusions

This work aimed to investigate using remote-power ultrasonic vibration to improve the laser welding process. Therefore, an efficient modelling approach combining the plate vibration method and the cavitation oscillation method was employed to analyse the forced vibration of cavitation in the welding pool. The proposed method provided an accurate tool for selecting the excitation position on the plate and the frequency range required to generate cavitation in the molten pool.

Time and frequency domain studies showed that for respective frequencies of 22, 32 and 40 kHz, the displacement reached at least a level of 0.1 pm, which can excite the semifluid in the fusion zone and burst the microbubbles.

The duration of the solidifying phase should be around 0.1 s, which corresponds to 200 cycles of ultrasound at 20 kHz. Theoretically, this is sufficient time to make the bubbles expand and burst in the semifluid. Hence, grain refinement and better shaping of intermetallic can be achieved.

Future study includes manufacturing laser-welded testing specimens under ultrasonic vibration and evaluating improvements in microstructure based on welding times and ultrasound power.

Author Contributions: T.-H.G. and A.T. developed the concept for this study; formal analysis was performed by A.T. and M.S.; original draft was written by A.T. and M.S., and reviewed and edited by E.E.M. and T.-H.G.; W.B. and T.-H.G. were responsible for supervision, investigation and project administration; T.-H.G. and. E.E.M. contributed to funding acquisition and to resource deployment. All authors have read and agreed to the published version of the manuscript.

Funding: This research was funded by Innovate UK, grant number 10018077. The AP Cwas funded by Brunel University London.

Institutional Review Board Statement: Not applicable.

Informed Consent Statement: Not applicable.

Data Availability Statement: Not applicable.

Conflicts of Interest: The authors declare no conflict of interest.

References

1. Wei, H.L.; Mazumder, J.; DebRoy, T. Evolution of Solidification Texture during Additive Manufacturing. *Sci. Rep.* **2015**, *5*, 16446. [CrossRef]
2. Franks, J.; Wheatley, G.; Zamani, P.; Masoudi Nejad, R.; Macek, W.; Branco, R.; Samadi, F. Fatigue Life Improvement Using Low Transformation Temperature Weld Material with Measurement of Residual Stress. *Int. J. Fatigue* **2022**, *164*, 107137. [CrossRef]
3. Zwicker, M.F.R.; Moghadam, M.; Zhang, W.; Nielsen, C.V. Automotive Battery Pack Manufacturing—A Review of Battery to Tab Joining. *J. Adv. Join. Process.* **2020**, *1*, 100017. [CrossRef]
4. Favez, D.; Wagniere, J.-D.; Rappaz, M. Au–Fe Alloy Solidification and Solid-State Transformations. *Acta Mater.* **2010**, *58*, 1016–1025. [CrossRef]
5. Ghanbari, H.R.; Shariati, M.; Sanati, E.; Masoudi Nejad, R. Effects of Spot Welded Parameters on Fatigue Behavior of Ferrite-Martensite Dual-Phase Steel and Hybrid Joints. *Eng. Fail. Anal.* **2022**, *134*, 106079. [CrossRef]
6. Mypati, O.; Sahu, S.; Pal, S.K.; Srirangam, P. An Investigation of Mechanical and Electrical Properties of Friction Stir Welded Al and Cu Busbar for Battery Pack Applications. *Mater. Chem. Phys.* **2022**, *287*, 126373. [CrossRef]
7. Li, Y.; Shen, K. The Effect of Melt Vibration on Polystyrene Melt Flowing Behavior during Extrusion. *J. Macromol. Sci. Part B* **2008**, *47*, 1228–1235. [CrossRef]
8. Bast, J.; Hübler, J.; Dommaschk, C. Influence of Vibration during Solidification of Molten Metals on Structure and Casting Properties. *Adv. Eng. Mater.* **2004**, *6*, 550–554. [CrossRef]
9. Nie, K.B.; Wang, X.J.; Wu, K.; Hu, X.S.; Zheng, M.Y.; Xu, L. Microstructure and Tensile Properties of Micro-SiC Particles Reinforced Magnesium Matrix Composites Produced by Semisolid Stirring Assisted Ultrasonic Vibration. *Mater. Sci. Eng. A* **2011**, *528*, 8709–8714. [CrossRef]
10. Huang, Z.; Lucas, M.; Adams, M.J. Influence of Ultrasonics on Upsetting of a Model Paste. *Ultrasonics* **2002**, *40*, 43–48. [CrossRef] [PubMed]
11. Prihandana, G.S.; Mahardika, M.; Hamdi, M.; Wong, Y.S.; Mitsui, K. Effect of Micro-Powder Suspension and Ultrasonic Vibration of Dielectric Fluid in Micro-EDM Processes—Taguchi Approach. *Int. J. Mach. Tools Manuf.* **2009**, *49*, 1035–1041. [CrossRef]

12. Jimma, T.; Kasuga, Y.; Iwaki, N.; Miyazawa, O.; Mori, E.; Ito, K.; Hatano, H. An Application of Ultrasonic Vibration to the Deep Drawing Process. *J. Mater. Process. Technol.* **1998**, *80*, 406–412. [CrossRef]
13. Khavari, M.; Priyadarshi, A.; Hurrell, A.; Pericleous, K.; Eskin, D.; Tzanakis, I. Characterization of Shock Waves in Power Ultrasound. *J. Fluid Mech.* **2021**, *915*, R3. [CrossRef]
14. Eskin, G.I. *Ultrasonic Treatment of Light Alloy Melts*; CRC Press: Boca Raton, FL, USA, 1998; ISBN 0429075863.
15. Apfel, R.E.; Holland, C.K. Gauging the Likelihood of Cavitation from Short-Pulse, Low-Duty Cycle Diagnostic Ultrasound. *Ultrasound Med. Biol.* **1991**, *17*, 179–185. [CrossRef]
16. Yasui, K. Influence of Ultrasonic Frequency on Multibubble Sonoluminescence. *J. Acoust. Soc. Am.* **2002**, *112*, 1405–1413. [CrossRef]
17. Brujan, E.A.; Ikeda, T.; Matsumoto, Y. On the Pressure of Cavitation Bubbles. *Exp. Fluid Sci.* **2008**, *32*, 1188–1191. [CrossRef]
18. Sun, Q.J.; Lin, S.B.; Yang, C.L.; Zhao, G.Q. Penetration Increase of AISI 304 Using Ultrasonic Assisted Tungsten Inert Gas Welding. *Sci. Technol. Weld. Join.* **2009**, *14*, 765–767. [CrossRef]
19. Chen, X.; Shen, Z.; Wang, J.; Chen, J.; Lei, Y.; Huang, Q. Effects of an Ultrasonically Excited TIG Arc on CLAM Steel Weld Joints. *Int. J. Adv. Manuf. Technol.* **2012**, *60*, 537–544. [CrossRef]
20. Lan, H.; Gong, X.; Zhang, S.; Wang, L.; Wang, B.; Nie, L. Ultrasonic Vibration Assisted Tungsten Inert Gas Welding of Dissimilar Metals 316L and L415. *Int. J. Miner. Metall. Mater.* **2020**, *27*, 943–953. [CrossRef]
21. Yang, F.; Zhou, J.; Ding, R. Ultrasonic Vibration Assisted Tungsten Inert Gas Welding of Dissimilar Magnesium Alloys. *J. Mater. Sci. Technol.* **2018**, *34*, 2240–2245. [CrossRef]
22. Zhou, S.; Ma, G.; Dongjiang, W.; Chai, D.; Lei, M. Ultrasonic Vibration Assisted Laser Welding of Nickel-Based Alloy and Austenite Stainless Steel. *J. Manuf. Process.* **2018**, *31*, 759–767. [CrossRef]
23. Liu, Z.; Jin, X.; Li, J.; Hao, Z.; Zhang, J. Numerical Simulation and Experimental Analysis on the Deformation and Residual Stress in Trailing Ultrasonic Vibration Assisted Laser Welding. *Adv. Eng. Softw.* **2022**, *172*, 103200. [CrossRef]
24. Liu, J.; Zhu, H.; Li, Z.; Cui, W.; Shi, Y. Effect of Ultrasonic Power on Porosity, Microstructure, Mechanical Properties of the Aluminum Alloy Joint by Ultrasonic Assisted Laser-MIG Hybrid Welding. *Opt. Laser Technol.* **2019**, *119*, 105619. [CrossRef]
25. Lei, Z.; Bi, J.; Li, P.; Guo, T.; Zhao, Y.; Zhang, D. Analysis on Welding Characteristics of Ultrasonic Assisted Laser Welding of AZ31B Magnesium Alloy. *Opt. Laser Technol.* **2018**, *105*, 15–22. [CrossRef]
26. Tarasov, S.Y.; Vorontsov, A.V.; Fortuna, S.V.; Rubtsov, V.E.; Krasnoveikin, V.A.; Kolubaev, E.A. Ultrasonic-Assisted Laser Welding on AISI 321 Stainless Steel. *Weld. World* **2019**, *63*, 875–886. [CrossRef]
27. Teyeb, A.; Silva, J.; Kanfoud, J.; Carr, P.; Gan, T.-H.; Balachandran, W. Improvements in the Microstructure and Mechanical Properties of Aluminium Alloys Using Ultrasonic-Assisted Laser Welding. *Metals* **2022**, *12*, 1041. [CrossRef]
28. Woizeschke, P.; Radel, T.; Nicolay, P.; Vollertsen, F. Laser Deep Penetration Welding of an Aluminum Alloy with Simultaneously Applied Vibrations. *Lasers Manuf. Mater. Process.* **2017**, *4*, 1–12. [CrossRef]
29. Radel, T. Thermal Impacts in Vibration-Assisted Laser Deep Penetration Welding of Aluminum. *Phys. Procedia* **2017**, *89*, 131–138. [CrossRef]
30. Yin, P.; Xu, C.; Pan, Q.; Zhang, W.; Jiang, X. Effect of Different Ultrasonic Power on the Properties of RHA Steel Welded Joints. *Materials* **2022**, *15*, 768. [CrossRef]
31. Wu, J.; Nyborg, W.L. Ultrasound, Cavitation Bubbles and Their Interaction with Cells. *Adv. Drug Deliv. Rev.* **2008**, *60*, 1103–1116. [CrossRef]
32. Prosperetti, A. Vapor Bubbles. *Annu. Rev. Fluid Mech.* **2017**, *49*, 221–248. [CrossRef]
33. Chladni, E.F.F. *Entdeckungen Über Die Theorie Des Klanges*; Zentralantiquariat der, Breitkopfund Ha¨rtel: Leipzig, Germany, 1787.
34. Leissa, A.W. *Vibration of Plates*; Scientific and Technical Information Division, National Aeronautics and Space Administration: Washington, DC, USA, 1969; Volume 160.
35. Sung, C.-C.; Chiu, C.-Y. Control of Sound Transmission through Thin Plate. *J. Sound Vib.* **1998**, *218*, 605–618. [CrossRef]
36. Sung, C.-C.; Jan, J.T. The Response of and Sound Power Radiated by a Clamped Rectangular Plate. *J. Sound Vib.* **1997**, *207*, 301–317. [CrossRef]
37. Sung, C.-C.; Jan, C.T. Active Control of Structurally Radiated Sound from Plates. *J. Acoust. Soc. Am.* **1997**, *102*, 370–381. [CrossRef]
38. Vlasov, V.S. *Some New Problems on Shells and Thin Structures*; Naca Tech: Washington, DC, USA, 1949.
39. Pinnington, R.J.; Briscoe, A.R. Externally Applied Sensor for Axisymmetric Waves in a Fluid Filled Pipe. *J. Sound Vib.* **1994**, *173*, 503–516. [CrossRef]
40. Louisnard, O.; González-García, J. Acoustic Cavitation. In *Ultrasound Technologies for Food and Bioprocessing*; Springer: Berlin/Heidelberg, Germany, 2011; pp. 13–64.
41. Jamshidi, R.; Brenner, G. Dissipation of Ultrasonic Wave Propagation in Bubbly Liquids Considering the Effect of Compressibility to the First Order of Acoustical Mach Number. *Ultrasonics* **2013**, *53*, 842–848. [CrossRef]
42. Brennen, C.E. *Cavitation and Bubble Dynamics*; Cambridge University Press: Cambridge, UK, 2014; ISBN 1107644763.

43. Ashokkumar, M. The Characterization of Acoustic Cavitation Bubbles—An Overview. *Ultrason Sonochem.* **2011**, *18*, 864–872. [CrossRef]
44. Chen, C.; Fan, C.; Cai, X.; Lin, S.; Liu, Z.; Fan, Q.; Yang, C. Investigation of Formation and Microstructure of Ti-6Al-4V Weld Bead during Pulse Ultrasound Assisted TIG Welding. *J. Manuf. Process.* **2019**, *46*, 241–247. [CrossRef]

materials

Article

An Improved Method for Deriving the Heat Source Model for FCAW of 9% Nickel Steel for Cryogenic Tanks

Younghyun Kim [1,2], Jaewoong Kim [1], Hyeongsam Park [3], Sungbin Hong [4], Changmin Pyo [1,*] and Gyuhae Park [2,*]

1 Automotive Materials & Components R&D Group, Korea Institute of Industrial Technology, Gwangju 61012, Republic of Korea; kyh1927@kitech.re.kr (Y.K.); kjw0607@kitech.re.kr (J.K.)
2 School of Mechanical Engineering, Chonnam National University, Gwangju 61186, Republic of Korea
3 Department of Computer Engineering, Chonnam National University, Gwangju 61186, Republic of Korea; 157485@jnu.ac.kr
4 Naval Architecture and Ocean Engineering, Seoul National University, Seoul 08826, Republic of Korea; njjo177@hshi.co.kr
* Correspondence: changmin@kitech.re.kr (C.P.); gpark@jnu.ac.kr (G.P.)

Abstract: The International Maritime Organization (IMO) is tightening regulations on air pollutants. Consequently, more LNG-powered ships are being used to adhere to the sulfur oxide regulations. Among the tank materials for storing LNG, 9% nickel steel is widely used for cryogenic tanks and containers due to its high cryogenic impact toughness and high yield strength. Hence, numerous studies have sought to predict 9% nickel steel welding distortion. Previously, a methodology to derive the optimal parameters constituting the Goldak welding heat source for arc welding was developed. This was achieved by integrating heat transfer finite element analysis and optimization algorithms. However, this process is time-consuming, and the resulting shape of the weld differs by ~15% from its actual size. Therefore, this study proposes a simplified model to reduce the analysis time required for the arc welding process. Moreover, a new objective function and temperature constraints are presented to derive a more sophisticated heat source model for arc welding. As a result, the analysis time was reduced by ~70% compared to that previously reported, and the error rates of the weld geometry and HAZ size were within 10% and 15% of the actual weld, respectively. The findings of this study provide a strategy to rapidly predict welding distortion in the field, which can inform the revision of welding guidelines and overall welded structure designs.

Keywords: flux core arc welding; 9% nickel steel (ASTM A553-1); Goldak welding heat source model; Evolutionary optimization algorithm; simplification model

Citation: Kim, Y.; Kim, J.; Park, H.; Hong, S.; Pyo, C.; Park, G. An Improved Method for Deriving the Heat Source Model for FCAW of 9% Nickel Steel for Cryogenic Tanks. *Materials* **2023**, *16*, 6647. https://doi.org/10.3390/ma16206647

Academic Editors: Swarup Bag and Christ Prakash Paul

Received: 24 August 2023
Revised: 5 October 2023
Accepted: 8 October 2023
Published: 11 October 2023

1. Introduction

As the effects of climate change accelerate, air pollution is becoming a major environmental problem. Therefore, the International Maritime Organization (IMO) has continued to tighten regulations on air pollutants, such as sulfur oxides (SO_x) and nitrogen oxides (NO_x), to reduce air pollution emanating from ships. From 2020, marine fuel's maximum permitted sulfur content has been reduced to 0.5% from 3.5% globally. Tier II regulations on nitrogen oxide (NO_x) emissions have been in effect since 2011. Meanwhile, Tier III regulations, which apply only to Emission Control Areas (ECA), have been in effect since 2016. Consequently, ships operating in ECAs must be equipped with certified Tier III engines [1–3].

The shipping industry developed measures to adapt to these regulations, including using low-sulfur fuel, installing scrubbers, and operating LNG-powered ships [4,5]. Among these measures, LNG-fueled ships are gaining increased attention from shipowners as they significantly reduce air pollutant emissions, including SO_x and NO_x. They are also more cost-effective than ships that use conventional fuels, which is advantageous regarding overall operating costs [6,7]. Compared to heavy fuel oils, liquefied natural gas (LNG) reduces the emission of NO_x by 85–95%, CO_2 by 20%, and SO_x by ~100%, enabling shipowners to comply with IMO environmental regulations [8,9]. When extracted, LNG fuel exists as natural gas and occupies a large volume. However, when cooled to temperatures below −163 °C (−261 °F), LNG undergoes liquefaction, and its volume is reduced to approximately 1/600th of its gaseous state [10–14]. Therefore, it is essential to develop tanks suitable for storing LNG fuel.

The IGC code specifies the materials that are suitable for use in storage tanks, including 304L stainless steel, 316L stainless steel, aluminum 5038-O, high manganese steel, 36% nickel steel, and 9% nickel steel, which have cryogenic toughness and a low thermal expansion coefficient. Among them, 9% nickel steel is widely used for producing the cryogenic equipment and containers of LNG ships [15]. However, certain issues persist, including high-temperature-induced cracking of the weld metal and heat-affected zone, and magnetism-induced arc blow during welding, a crucial process in cryogenic equipment and container production [16]. Hence, many studies have sought to improve the 9% nickel steel welding efficiency. For instance, Kim et al. performed butt welding via shielded metal arc welding (SMAW), submerged arc welding (SAW), and flux-cored arc welding (FCAW) on 9% nickel steel. They then performed impact testing at cryogenic temperature (−196 °C) and compared the results [16]. Meanwhile, Pyo conducted BOP tests on four shielding gases using fiber laser welding with 9% nickel steel and compared the penetration and HAZ depth through cross-sectional analysis [17]. Park et al. performed butt welding with super-TIG using Alloy625 filler metal and FCAW using Alloy600 filler metal on 9% nickel steel. They then compared the welds by conducting mechanical tests [18]. Xu et al. applied deep penetration keyhole tungsten inert gas (K-TIG) welding on 9% nickel steel. They analyzed the relationship between the grain size of martensite and the mechanical properties of the weld via cryogenic (−196 °C) impact tests and microstructure investigation [19]. Gook et al. performed analyses on Ytterbium fiber laser welding after preheating 9% nickel steel. Through cryogenic (−196 °C) impact testing, they established the optimal preheating temperature range to produce a sufficiently high-impact toughness comparable to the base metal [20].

FCAW has exhibited high efficiency as it has been used to develop welded wires with toughness and strength equivalent to the base metal. In turn, FCAW research is underway on cryogenic materials. For instance, Kim performed FCAW using hot steroid series DW-709SP filler metal on 9% nickel steel and achieved weld reliability through tensile, bending, hardness, and cryogenic (−193 °C) impact testing, as required for WPS approval [21]. Meanwhile, Mu et al. used DW-N625 filler metal on 9% nickel steel and performed crack tip opening displacement (CTOD) testing at room temperature (23 °C) and cryogenic temperature (−193 °C); they also observed the microstructure microscopically. Precipitation from TCP and carbides was observed on the weld, confirming that the number of precipitates depending on the welding conditions affected the fracture toughness. They also found that crack propagation caused by precipitation often occurs at cryogenic temperatures [22]. Additionally, Park et al. measured the bead geometry, area, and hardness of welds created via FCAW on 9% nickel steel. They found that the hardness varies depending on the ratio of wires mixed in the weld. Moreover, the weld quality decreases when the mixing ratio is ≥15.0%; the optimal welding conditions were derived using the multi-objective optimization (MOO) algorithm [23]. Moshtaghi et al. studied the FCAW and SMAW welding behavior of Ni addition ferritic weldments, confirming that applying FCAW increases the density of high-angle grain boundaries compared to SMAW [24].

Welding involves melting a localized area of material. The weld zone rapidly reaches a high temperature and then rapidly cools through heat transfer to a low-temperature area with a large volume. This drastic temperature change alters the material's mechanical properties and causes residual stress and distortion of the welding, resulting in increased cost and time spent due to the need to select the proper welding conditions and equipment. In addition, the strength and dimensional accuracy of the welded structure is reduced.

Therefore, many analytical studies have been conducted to reduce the required trial and error through finite element analysis. For example, Kim et al. conducted a 2D finite element analysis to investigate the residual stress and temperature distribution while welding 9% nickel steel plates. The finite element analysis results were consistent with the actual weld geometry and temperature distribution; the resulting residual stress was most apparent in the longitudinal direction of the weld line [25]. Manurung et al. performed multipass GMAW on a T-shaped structure comprising low manganese carbon steel and compared the weld temperature distribution and thermal distortion with 3D finite element analysis [26].

Meanwhile, Deng et al. performed CO_2 welding on a thin panel structure made of 570 MPa thermo-mechanical control press (TMCP) steel. They then developed a welding distortion prediction method by comparing and analyzing the effects of thermal distortion, welding heat input, and sequence through 3D finite element analysis [27]. García-García et al. performed butt welding of 5.8 mm twinning-induced plasticity (TWIP) steel plates via the GTAW process. The temperature distribution according to the heat input was confirmed through 3D finite element analysis and was compared with the microstructure results of the welded specimens [28]. Moreover, Pyo et al. estimated the parameters of the arc heat source based on the results of applying FCAW to 9% nickel steel [29]. Collectively, this research was performed to establish a methodology to derive the main parameters constituting the welding heat source for each welding condition using bead-on-plate experiments, heat transfer finite element analysis, and optimization algorithms. However, the global optimization algorithm had to perform the time-consuming process of comparing 2000 candidates, while the size of the actual fusion zone and the analysis result differed by ~14%. This difference was caused by controlling only the heat-affected zone (HAZ) size through the objective function and not the fusion zone size.

Therefore, this study proposes a simplified model to address the slow analysis speed and presents precise objective functions and constraints to derive an arc welding heat source model comprising the fusion zone and HAZ sizes. Based on this model, the main parameters of the heat source were derived for three welding conditions, including those from the previous study. The derived heat source and weld geometry were determined experimentally and subsequently compared and analyzed; the differences between the simplified analysis model and the 3D-based full model were also compared. The simplified analysis model reduced analysis time by >70% compared to the previous study. Moreover, the error rates of the weld geometry and HAZ size compared to the actual weld were within 10% and 15%, respectively. Deriving an arc welding heat source model with a sophisticated fusion zone facilitated the prediction of welding distortion that occurs during arc welding with a small error rate. Hence, the findings of this study may help establish guidelines for welding in the field with 9% nickel steel and evaluate the changes in the welding location and structure design in consideration of the amount of distortion caused by thermal contraction or expansion. Given that this method uses analytics to predict welding distortion, the associated cost will be reduced as it will facilitate troubleshooting within the design stage, avoiding the requirement for actual experiments.

2. Welding Experiments and Results

2.1. Welding Materials and Conditions

This study used 9% nickel steel (ASTM A553-1) [30], and bead-on-plate (BOP) welding was performed using the FCAW process. The dimensions of the test specimen plates were 150 mm (W) × 200 mm (L) and were 15 mm thick. Welding was performed by fixing the ends of the four corners. Figure 1 shows a schematic diagram, including the welding direction. The wire used in the experiment was AWS A5.14 ERNiMo-8 (KOBELCO, TG-S709S, Changwon-si, Korea) with a diameter of 1.2 mm. Table 1 shows the chemical composition of ASTM A553-1 [30], and Table 2 presents its mechanical properties [30].

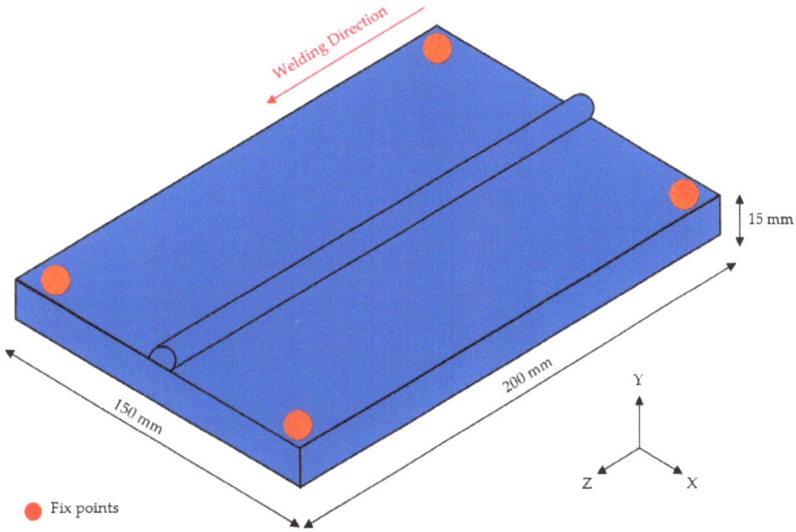

Figure 1. Schematic diagram of welding.

Table 1. Chemical composition of base metal and filler wire (wt.%) [30].

	C	Si	Mn	S	P	Ni	Fe
Parent material	0.05	0.67	0.004	0.003	0.25	9.02	Bal.
Welding consumables	0.02	0.02	0.1	0.001	0.001	69.8	Bal.

Table 2. Mechanical properties of 9% nickel steel [30].

Yield Strength (MPa)	Tensile Strength (MPa)	Elongation (%)	Hardness (HV)
651.6	701.1	26.6	243

Welding conditions included current, voltage, and welding speed, the main variables that affect heat input. The voltage was fixed at 15 V and the welding speed at 0.4 m/min; the experiment was conducted by changing only the current. Table 3 shows the welding conditions, including the shielding gas.

Table 3. FCAW parameters and experimental condition.

Case	Current (A)	Voltage (V)	Welding Speed (m/min)	Shielding Gas (L/min)
Case 1	150	25	0.4	18
Case 2	160	25	0.4	18
Case 3	170	25	0.4	18

2.2. Cross-Section Analysis Results

FCAW welding was performed with a 600 A class welding machine (ProPAC, HYOSUNG, Mapo-gu, Seoul, Republic of Korea) comprising a torch, weld feeder, direct welding carriage, and rail. After BOP welding with the FCAW equipment, the specimens were cut transversely at the center point of the weld line to observe the weld geometry. The cut sections were sprayed with 90% ethanol and 10% nitric acid, and the shape of the weld was analyzed using an optical microscope (EGVM 35B, EG Tech, Anyang, Republic of

Korea) to measure the main welding parameters (bead width, bead height, HAZ width, and HAZ height). Figure 2 shows the parameter measurement locations for each case, and Table 4 and Figure 3 shows the measurements. As a result of cross-sectional analysis, it was confirmed that the measurement parameters increased every time the current value increased by 10A. Bead Height increased by ~6.51%, bead width by ~18.00%, HAZ depth by ~12.39%, and HAZ width by ~9.73%. The size of the fusion zone and HAZ area increased as the magnitude of the current increased. The difference might be due to the increase in heat input.

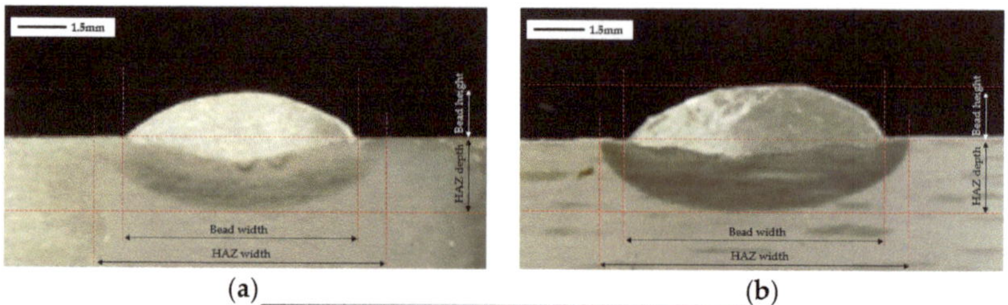

Figure 2. (**a**) Case 1, (**b**) Case 2, (**c**) Case 3, Parameter measurement location for each.

Table 4. Key parameter values.

Case	Bead Height (mm)	Bead Width (mm)	HAZ Depth (mm)	HAZ Width (mm)
Case 1	2.90	9.68	4.14	15.74
Case 2	3.09	11.35	4.66	16.64
Case 3	3.29	13.48	5.23	18.92

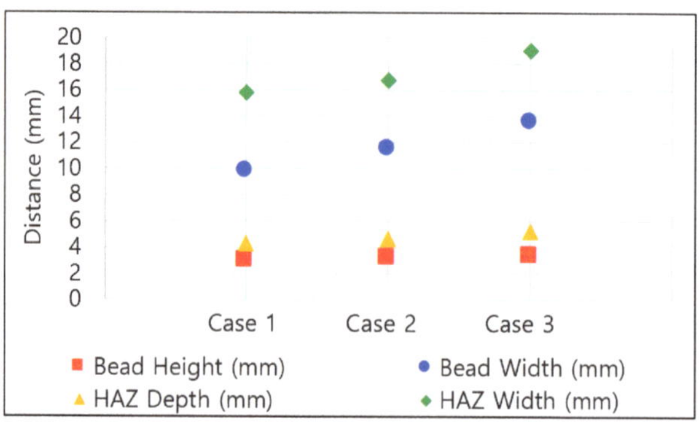

Figure 3. Cross-sectional analysis results by case.

3. Deriving the Heat Source Model

3.1. Process for Deriving the Heat Source

The process comprised BOP welding, measuring the weld geometry, and deriving the heat source results. Abaqus (Ver. 2020, Dassault Systems Simulia Corp, Johnston, RI, USA) was used as the finite element analysis program for heat transfer analysis. Considering the previously reported high reliability of Fortran user subroutines [31–35], they were used to implement the moving heat source (Fortan Ver. 17.0, Intel, San Jose, CA, USA).

Welding heat transfer analysis requires temperature-specific mechanical properties of the material, especially data at high temperatures. As the Jmatpro program has been used to derive the mechanical properties of metals [36,37], it was employed in this study to derive temperature-specific physical properties of 9% nickel steel. In addition, heat transfer analysis was performed by applying physical property data according to temperature to the analysis model. Figure 4 shows the thermal conductivity, specific heat, and density by temperature (25–2500 °C). A sharp decrease in density values was observed around the melting temperature of approximately 1500 °C for 9% nickel steel.

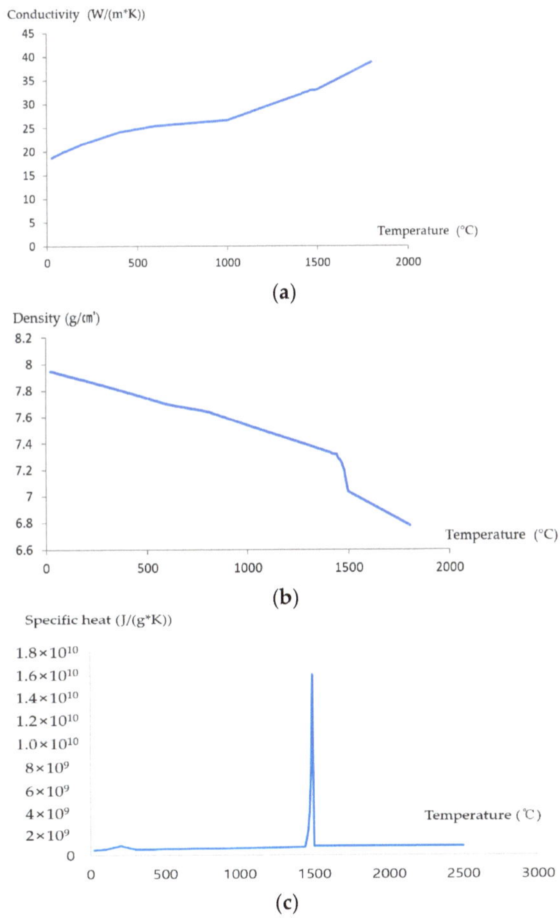

Figure 4. Material properties of 9% nickel steel by temperature (**a**) Conductivity, (**b**) Density, (**c**) Specific heat.

A finite element analysis model was constructed based on the experiments, and a moving heat source was applied to the center of the 150 mm wide, 200 mm long, and 15 mm thick sample to analyze the welding process. For 3D analysis, the mesh size was set to 2 mm, and ~90,000 meshes were used. Its accuracy was lower than a 2D model with a mesh size of 0.8 mm. Hence, as more than 2000 comparative analyses are required in this study, the 2D shell model was simplified to shorten the analysis time (Figure 5). The simplified model had a DC2D4 type mesh with a 4-node linear heat transfer quadrilateral and ~7000 meshes. The analysis quality was validated using a previously reported method [38]. The optimal arc heat source derived by the algorithm was applied to the 3D and 2D models to perform heat transfer analysis and compare the results. The convective heat transfer coefficient was set to 10 W/m^2K, the emissivity to 0.8, and the air temperature to 20 °C [29].

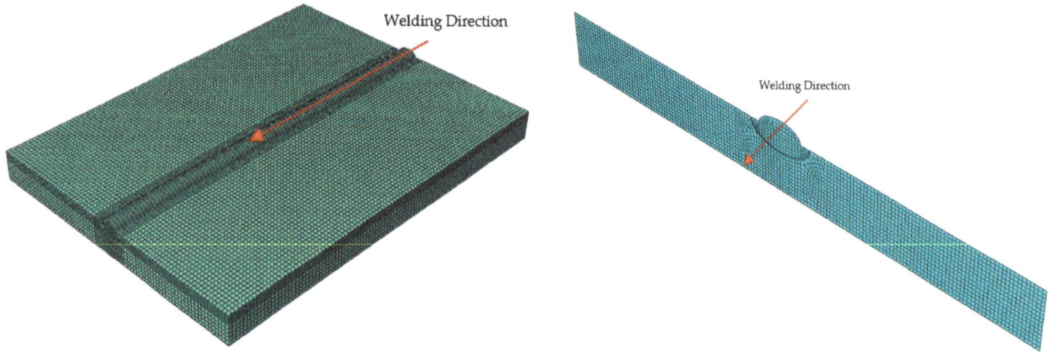

(**a**) (**b**)

Figure 5. Original dimension model and simplified model (**a**) Original dimension model (with coarse meshes), (**b**) Simplified model (with fine meshes).

3.2. Simplifying the Heat Transfer Analysis Model

Due to the difficulty associated with conducting experiments in the field, welding heat transfer analysis is often used to identify trends. However, the heat transfer analysis performed in previous studies are time intensive due to the large number of meshes, impeding the ability to rapidly identify trends in the field. Our previous study used a model with a 0.15 mm mesh and ~9000 meshes. However, in this study, the mesh size was reduced to 0.80 mm, and the count was increased to ~1200 to increase the analysis speed. Moreover, the model was simplified by ~80% compared to the previous study to compare and analyze the temperature distribution of the weld. As a result, the error rate of the weld geometry was ~20%, and the error rate of the HAZ size was ~7% compared to the previous experimental values. Meanwhile, in this study, the weld geometry and the HAZ size error rates were within 10% and 15% of the experimental values, respectively. Moreover, the analysis time decreased by more than 70%. More specifically, the previous model required 1680 s to analyze one model while the simplified model required 420 s. Therefore, applying a simplified model was considered more appropriate to identify trends and was used in the current study to derive the heat source. Figure 6 compares the simplified and previous models.

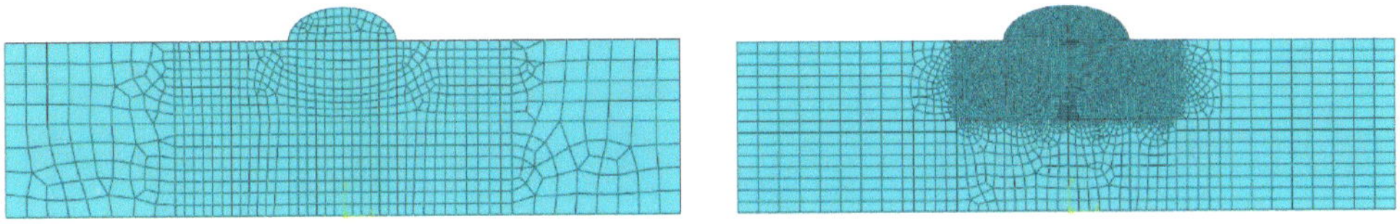

(**a**) (**b**)

Figure 6. Mesh shape by zone (**a**) Simplification model, (**b**) The previous study model.

3.3. Goldak Model Heat Source

The Goldak double ellipsoid model (Figure 7) is the most used heat source model in finite element analysis of arc welding, defined by the governing Equations (1)–(5). In Equations (2) and (3), the parameters a_f, a_r, b, and c are independent so that they can have different values for the front and back heat sources. Appropriate parameter values are required to simulate a geometry similar to the actual heat source; hence, many studies have sought to estimate these values. Farias et al. proposed a new heat source geometry based on Goldak's double ellipsoid model and applied a genetic algorithm to estimate the heat source for arc welding [39,40]. Meanwhile, Pyo used adaptive simulated annealing (ASA) to estimate the arc welding heat source [41].

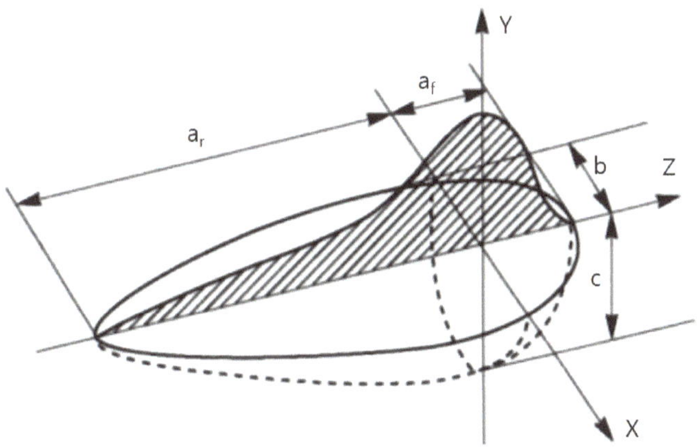

Figure 7. Heat distribution of Goldak model.

$$Q = \mu VI \tag{1}$$

$$q_f(x,y,z,t) = \frac{6\sqrt{3}f_f Q}{a_f bc\pi} \exp\left(-3\left(\frac{(z-vt-z_0)^2}{a_f{}^2} + \frac{y^2}{c^2} + \frac{x^2}{b}\right)\right) \tag{2}$$

$$q_r(x,y,z,t) = \frac{6\sqrt{3}f_r Q}{a_r bc\pi} \exp\left(-3\left(\frac{(z-vt-z_0)^2}{a_r{}^2} + \frac{y^2}{c^2} + \frac{x^2}{b^2}\right)\right) \tag{3}$$

$$f_f = \frac{2a_r}{a_f + a_r} \tag{4}$$

$$f_r = \frac{2a_f}{a_f + a_r} \tag{5}$$

Q: Effective heat energy
μ: Welding efficiency
V: Voltage
I: Current
f_f: Fraction of heat deposited at the front ellipsoid
f_r: Fraction of heat deposited at the rear ellipsoid
v: Velocity of heat source
a_f, a_r, b, c: Dimension parameters (Shown in Figure 7)

4. Optimization Algorithm

4.1. *Software*

The optimization algorithm program used in this study was Isight (Ver. 2020, Dassault Systems Simulia Corp, Johnston, RI, USA), designed by the same company that created Abaqus. Many studies have been conducted to design optimization algorithms using Isight [42–45].

4.2. *Algorithm Process*

Heat transfer analysis was repeated by altering six parameters, including the welding efficiency and the Goldak model parameters. It was, therefore, necessary to determine the parameter value with the closest temperature distribution based on comparing the analysis results with the actual HAZ size in Table 4. To find the optimal parameters of the arc heat source, the evolutionary optimization algorithm (Evol) method was applied (Figure 8) [46]. Evol is an evolutionary algorithm-based method that randomly adds normally distributed values to each design variable and applies mutation (standard deviation of the normal distribution) to identify the optimal value. The parental generation is randomly selected, and mutation operations are used to generate offspring. After evaluating the results produced by the mutations, the most suitable results replace the parents of the next generation; this process is repeated to converge with the objective function. In this way, Evol can solve problems with nonlinear constraints as it does not require any information other than that derived from the object evaluation method. As such, it has been used in many optimization processes [47,48].

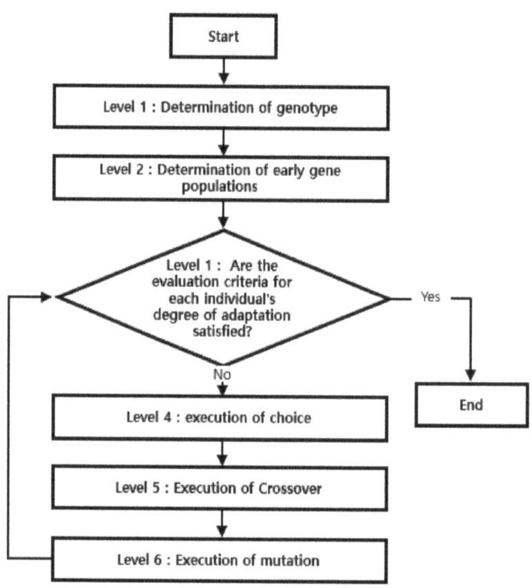

Figure 8. Evolutionary optimization algorithm [49].

4.3. Limiting Parameter Temperature Range

Based on the test results in Table 4, a 1.0 mm offset between the weld bead and HAZ boundary and inside and outside the HAZ border was applied to the model. In the previous study, temperatures were checked at five points at the offsets inside ($Q_{1–5}$) and outside ($P_{1–5}$) the HAZ; however, the temperature at the fusion zone was not recorded. Consequently, it was impossible to control or accurately predict the size of the fusion zone. Therefore, in this study's analysis model, the temperatures were checked at two weld points ($W_{1–2}$), five points at the offsets inside ($Q_{1–5}$) and outside ($P_{1–5}$) the HAZ, and five points at the HAZ boundary ($T_{1–5}$) (Figure 9). The weld points were at locations completely melted in the cross-sectional analysis (Figure 2), the end of the bead, and where the centerline of the bead met the base metal. Adding weld points reduces the range in finding an appropriate solution for the multivariate function and helps derive an arc welding heat source with a precise weld.

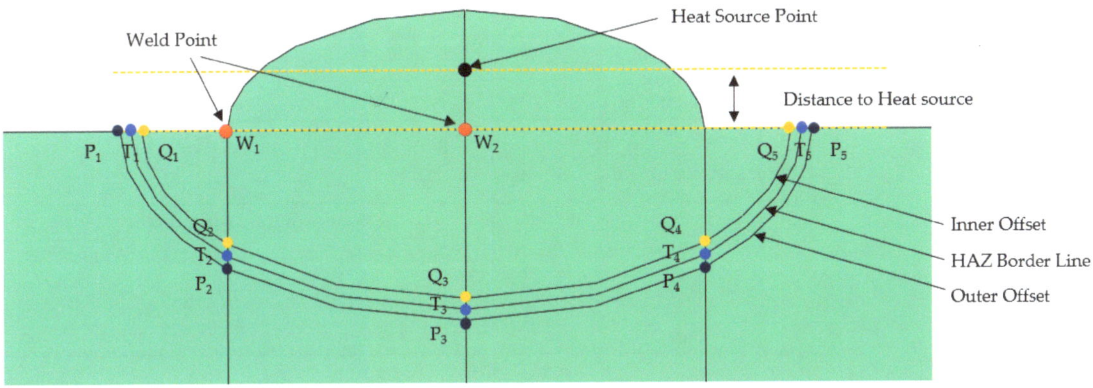

Figure 9. Offset shape, temperature checking position, definition of distance to heat source.

The following temperature constraints were assigned to derive a highly consistent heat source through the optimization algorithm: the weld point must exceed 1450 °C and the inner offset point must reach 600 °C at least once, while the outer offset point cannot exceed 600 °C. The temperatures of both the inner and outer offset constraints were set to 600 °C as carbon steel has an Ac1 point where its structure changes at temperatures above 600 °C. Therefore, 600 °C was selected as the temperature that separates the HAZ, and 1450 °C was applied for the fusion zone as melting occurs above this temperature. In the previous study, the sum of the differences between the maximum values of the inner and outer points was set as the objective function. However, as the offset interval widens, the range of the objective function value widens and becomes less consistent. Therefore, in this study, the HAZ boundary point was included in the objective function, and Equation (6) was applied by setting the value obtained by subtracting 600 °C from the sum of the maximum values as the objective function, as follows:

$$\text{Objective funtion} = \text{Min} \sum_{n=1}^{k} |T_{Tn} - h| \tag{6}$$

T_{Tn}: Temperature of HAZ Border Line check point n
n: Number of check point (n = 1, 2, 3, 4, 5)
k: Total number of check points (k = 5)
h: HAZ boundary temperature

4.4. Setting Variables and Ranges

This study sought to identify the optimal values of the parameters that minimize the objective function while satisfying the temperature constraints. There were six parameters in total. The parameters of the Goldak model, welding efficiency, and distance from the heat source were set as variables. As shown in Table 5, the upper bound was increased to find an optimal solution within a wide range. The welding efficiency was set to 5% (±2.5%) with 0.8 as the reference value. The variables a_f, a_r/a_f, b, and c were set to a relatively wide range. The distance to the heat source range was within the bead stacked on the upper part of the base metal, and Table 6 shows the values set as the bead's height for each case in Table 4. Additionally, Figure 9 presents the distance to heat source locations and ranges.

Table 5. Variables and ranges.

Variable	Lower Bound	Upper Bound
μ (W/W)	0.78	0.82
a_f (mm)	1.0	15.0
a_r/a_f (mm/mm)	1.5	7.0
b (mm)	1.0	20.0
c (mm)	1.0	15.0

Table 6. Distance to heat source ranges.

	L (Distance to Heat Source, mm)	
	Lower Bound	Upper Bound
Case 1	0	2.90
Case 2	0	3.09
Case 3	0	3.29

5. Results and Analysis

5.1. Deriving the Heat Source Parameters and Heat Transfer Analysis Results

Approximately 2000 candidates were compared using Isight. The parameter with the smallest objective function was derived and selected as the optimal value. Table 7 shows the derived parameter values, all were within the bound range established in Table 5. In addition, Table 7 shows a comparison between parameters from previous studies.

Table 7. Derived heat source parameters.

Variable	Value		
	Case 1	Case 2	Case 3
μ (W/W)	0.82	0.81	0.82
a_f (mm)	2.96	1.84	7.72
a_r/a_f (mm)	5.79	6.56	7
b (mm)	13.92	19.62	11.26
c (mm)	4.64	1.84	1.00
L (mm)	2.09	3.09	3.29

The welding heat input differed for each case (Table 3). The a_r and c parameters, constituting the heat source, showed a tendency, while a_f and b did not. This might have been caused by the offset range of 1.0 mm. Hence, the correlation between the heat input amount and the heat source parameter cannot be shown.

Navid et al. derived the optimal parameters of the Goldak heat source using the artificial neural network (ANN) method and compared the actual experimental results with the simulation results. They confirmed that the cross-section of the actual welding and simulation results were similar. However, the optimal parameters of the Goldak heat source derived from the ANN were also difficult to correlate with welding conditions [50].

Table 8 shows the values at the temperature constraint points obtained from the heat transfer analysis. Under all conditions, the maximum temperature at the outer offset $P_{1\text{-}3}$ points did not exceed 600 °C, while those at all of the inner offset $Q_{1\text{-}3}$ points were above 600 °C. The maximum temperature at the weld point was >1450 °C, and the temperature was ~600 °C at the HAZ boundary.

Table 8. Maximum temperature at each checkpoint per welding condition.

Temperature (°C)	Value		
	Case 1	Case 2	Case 3
P_1	546.83	522.86	266.03
P_2	413.66	389.53	407.15
P_3	419.67	391.85	458.52
Q_1	877.38	684.56	643.07
Q_2	787.20	830.43	821.43
Q_3	876.75	909.68	994.34
M_1	1450.84	1701.07	1474.87
M_2	2117.31	2692.81	3302.30
T_1	705.26	596.45	425.85
T_2	586.05	586.16	592.86
T_3	618.90	612.57	691.02

5.2. Heat Transfer Analysis Results by Welding Conditions

Figures 10–12 show the results of analyzing heat transfer in each case as time elapsed. The results were then compared from 0.1 s before the heat source arrived to 4.0 s after the heat source passed. In addition, the maximum HAZ width and HAZ depth were measured to compare the experimental and analytical results and obtain the error rate. The maximum HAZ width of Cases 1 and 2 was reached the instant the heat source arrived, and the maximum HAZ depth was reached 4.0 s after the heat source passed. The maximum HAZ width in Case 3 also was reached 1.0 s after the heat source arrived, and the maximum HAZ depth was reached 4.0 s after the heat source passed.

The maximum HAZ width and depth were measured from the heat transfer analysis results applying the simplified model for each case, and compared with the actual experimental results in Table 4 (Table 9).

Table 9. HAZ dimension and comparison.

Value	HAZ Width			HAZ Depth		
	FEM (mm)	Experiment (mm)	Difference (%)	FEM (mm)	Experiment (mm)	Difference (%)
Case 1	16.05	15.74	1.96	4.75	4.14	14.73
Case 2	15.84	16.64	4.81	5.20	4.66	11.59
Case 3	18.50	18.92	2.22	5.77	5.23	10.33

Case 1 had the smallest HAZ width error rate (1.96%), however, all cases had error rates within 4.81% and HAZ depth error rates within 14.73%. Hence, the 1.0 mm offset range significantly affects the HAZ depth measurement, which is relatively small compared to the HAZ width measurement. To find a more precise heat source model, reducing the offset range may reduce the error rate.

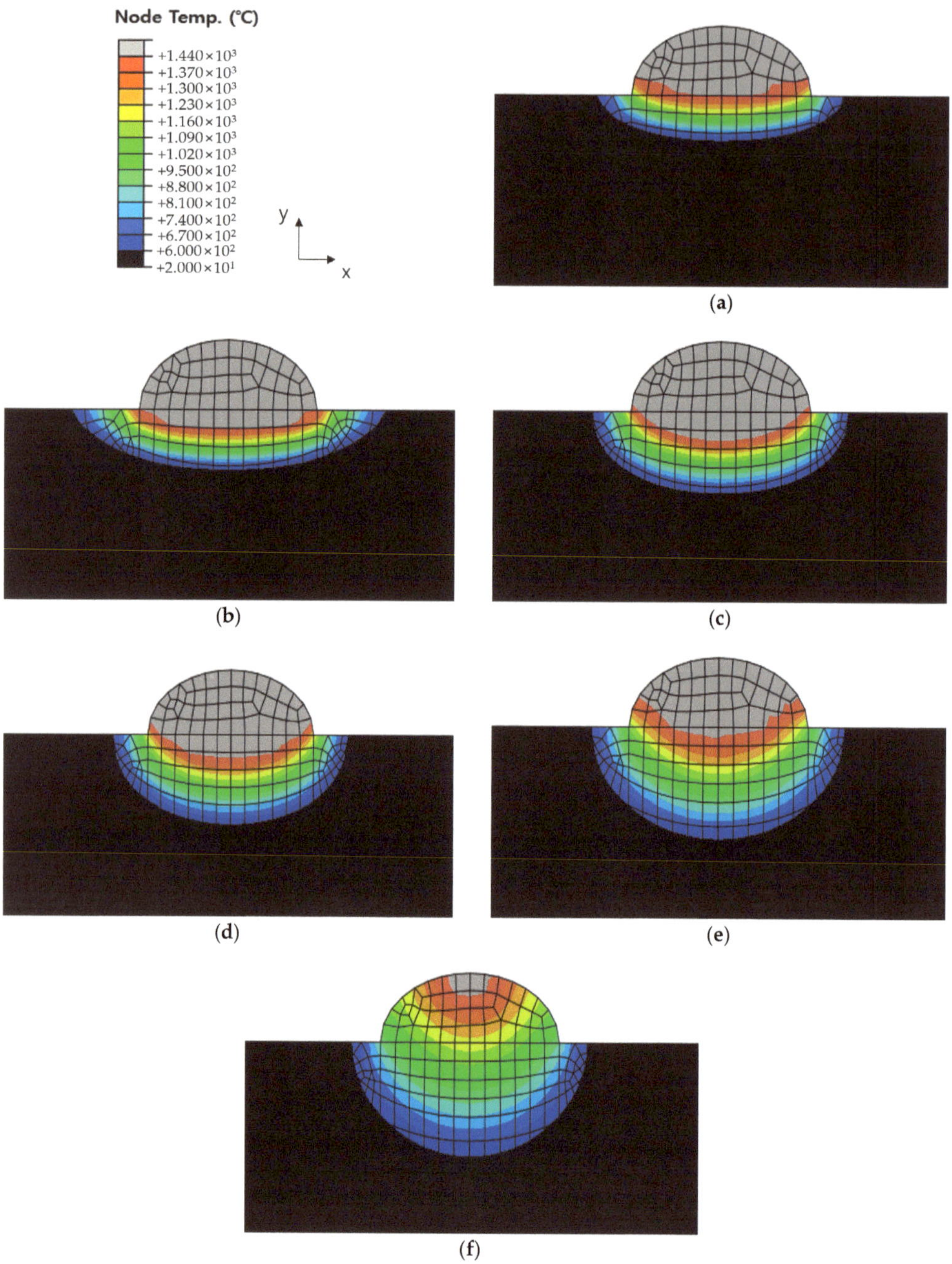

Figure 10. Temperature distribution by timeline before and after welding of Case 1. Times were (**a**) −0.1 s (before the heat source arrives); (**b**) 0 s (when the heat source is directly above); (**c**) 0.5 s elapsed; (**d**) 1.0 s elapsed; (**e**) 2.0 s elapsed; (**f**) 4.0 s elapsed.

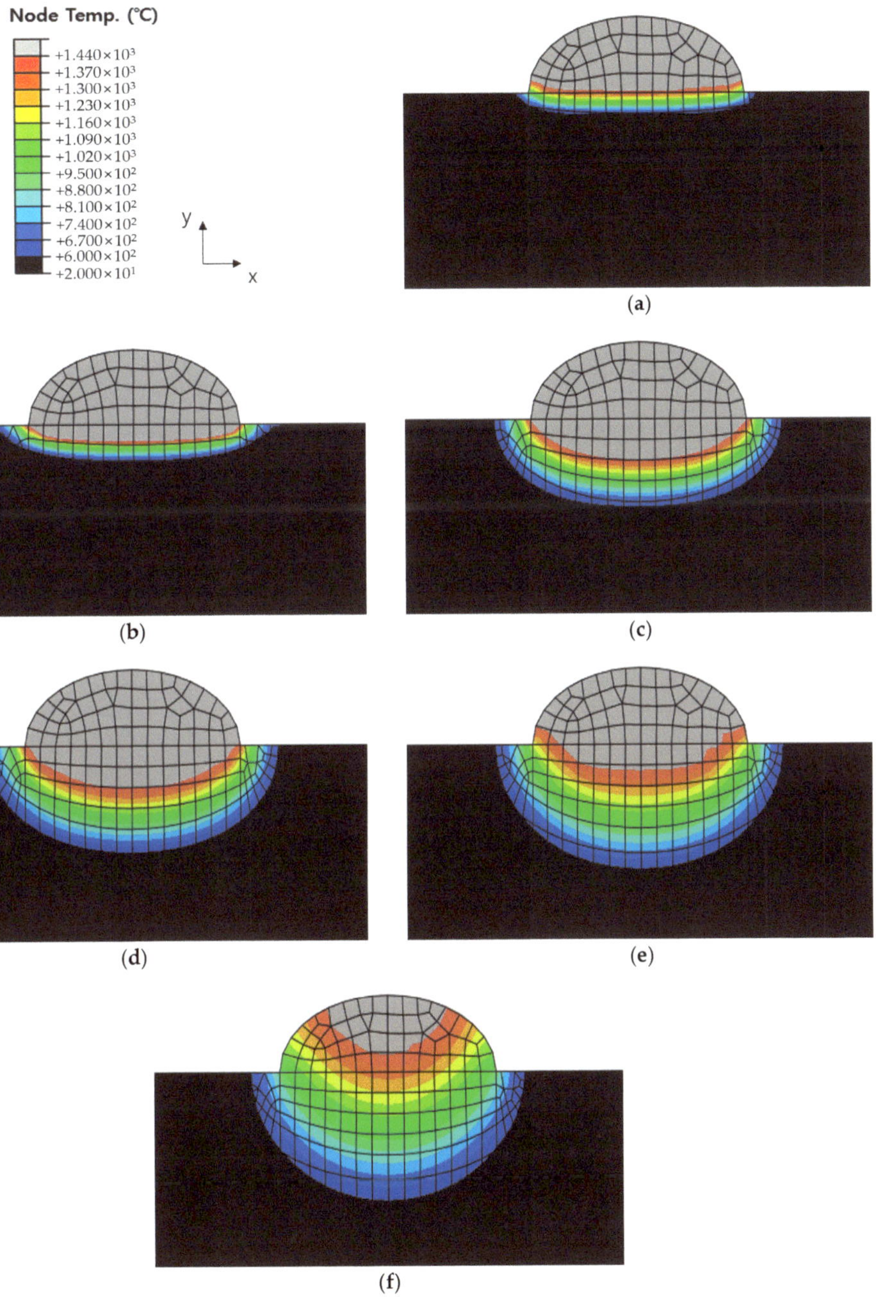

Figure 11. Temperature distribution by timeline before and after welding of Case 2. Times were (**a**) −0.1 s (before the heat source arrives); (**b**) 0 s (when the heat source is directly above); (**c**) 0.5 s elapsed; (**d**) 1.0 s elapsed; (**e**) 2.0 s elapsed; (**f**) 4.0 s elapsed.

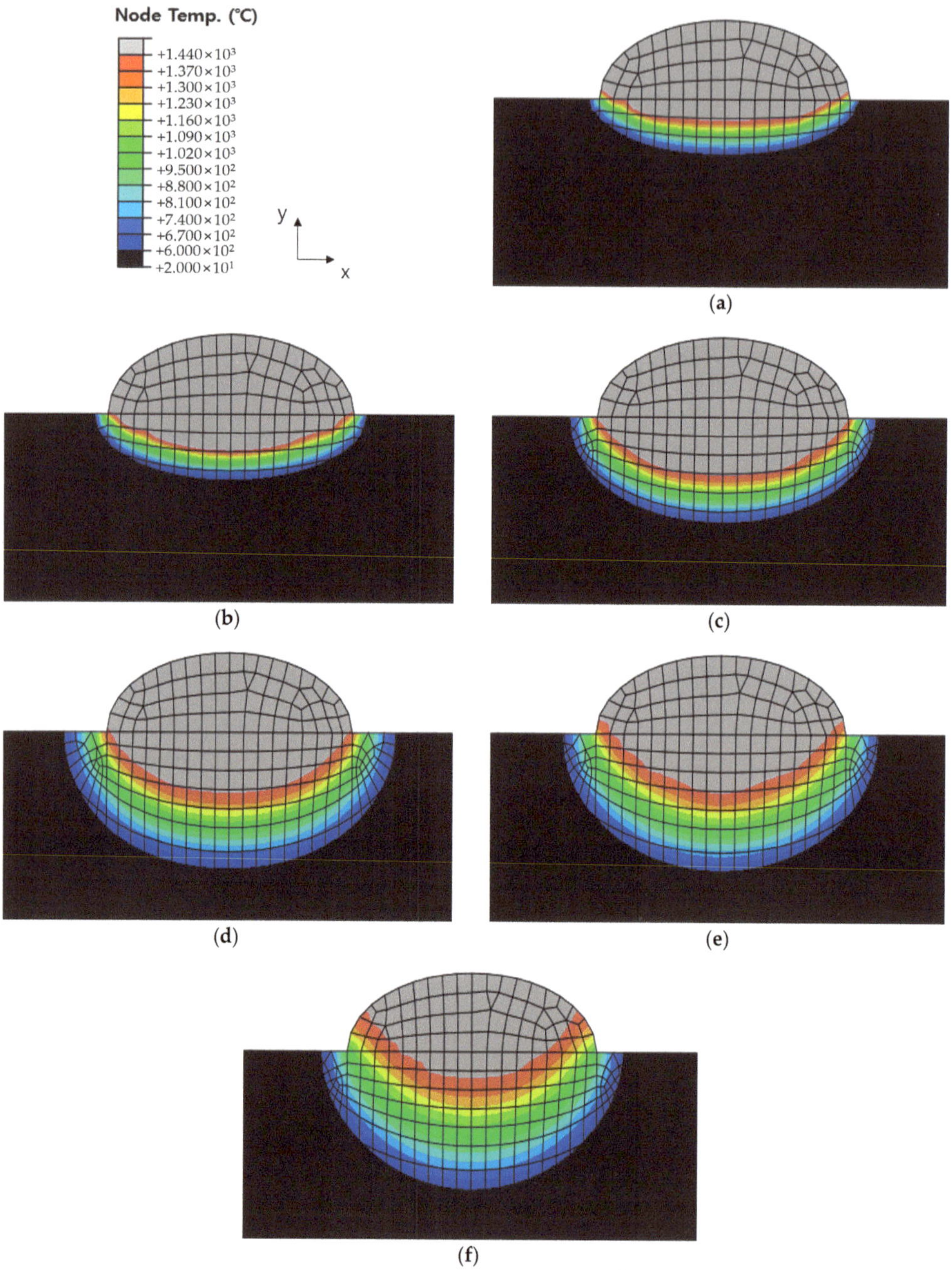

Figure 12. Temperature distribution by timeline before and after welding of Case 3. Times were (**a**) −0.1 s (before the heat source arrives); (**b**) 0 s (when the heat source is directly above); (**c**) 0.5 s elapsed; (**d**) 1.0 s elapsed; (**e**) 2.0 s elapsed; (**f**) 4.0 s elapsed.

5.3. Comparing the Heat Transfer Results between the Original Dimension Model and the Simplified Model

The original dimension model was simplified, as shown in Figures 5 and 6, and the optimal heat source parameter values were derived by applying the optimization algorithm and objective function to the simplified model. Therefore, it is necessary to compare and verify the original dimension model and the simplified model. The default mesh size of the original dimension model was set to 2 mm and 1 mm within 10 mm on both sides from the center of the weld. The DC3D8R element was used as the mesh type; hence, an 8-node linear heat transfer brick and reduced integration were applied. The mesh count for Cases 1, 2, and 3 were 90,300, 90,900, and 91,900, respectively. The temperature distribution was checked 1.6 s after the heat source passed at 18 mm, 86 mm, and 186 mm of the original dimension model and compared with the simplified model. Figure 13 shows the positions for checking the temperature distribution. Figures 14–16 show the comparison of Cases 1, 2, and 3 results, respectively (4-node linear heat transfer quadrilateral).

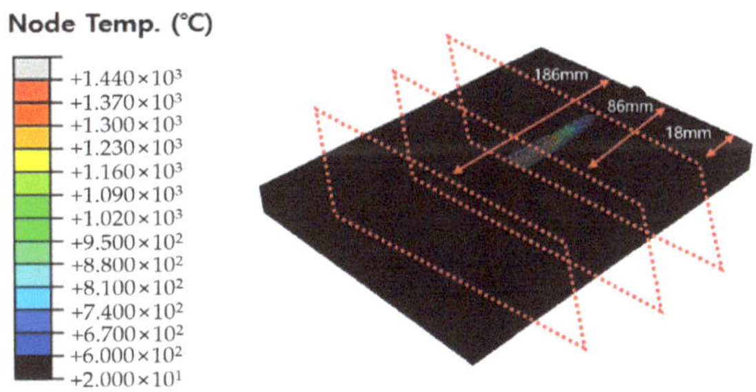

Figure 13. Section check position in the heat transfer analysis for the original dimension model.

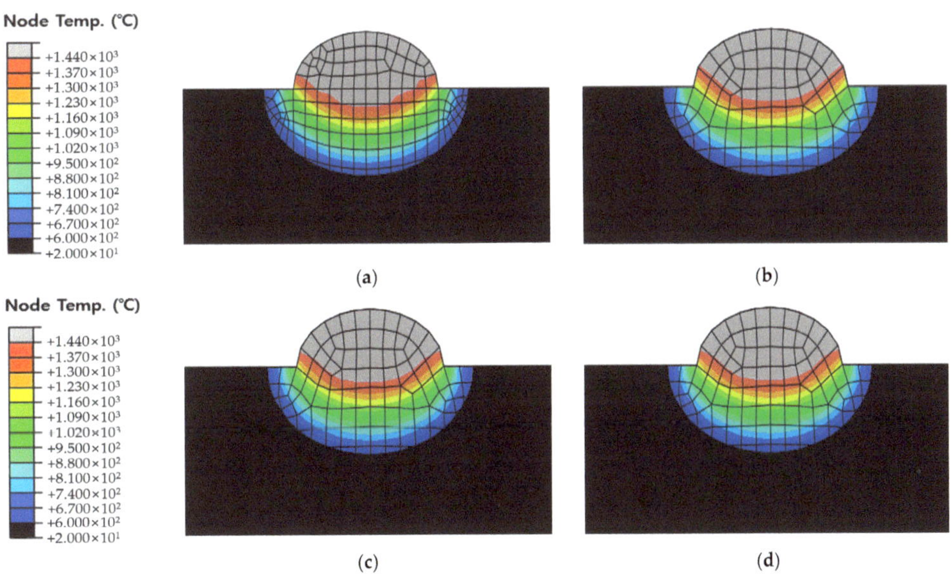

Figure 14. Comparison of temperature distribution of Case 1 and Original dimension model 1.6 s after the heat source has passed (**a**) Simplified model with fine meshes; (**b**) 18 mm position of original dimension model; (**c**) 86 mm position of original dimension model; (**d**) 186 mm position of original dimension model.

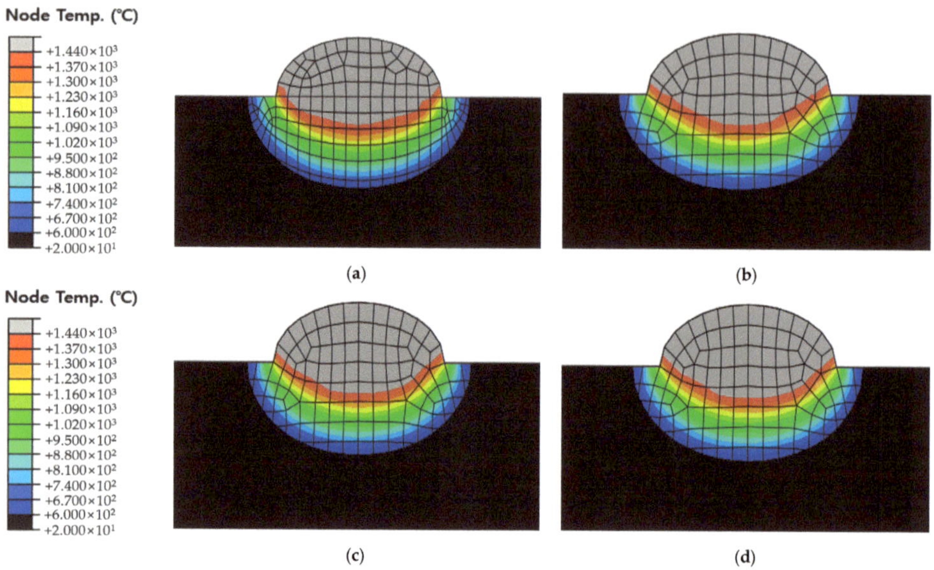

Figure 15. Comparison of temperature distribution of Case 2 and Original dimension model 1.6 s after the heat source has passed (**a**) Simplified model with fine meshes; (**b**) 18 mm position of original dimension model; (**c**) 86 mm position of original dimension model; (**d**) 186 mm position of original dimension model.

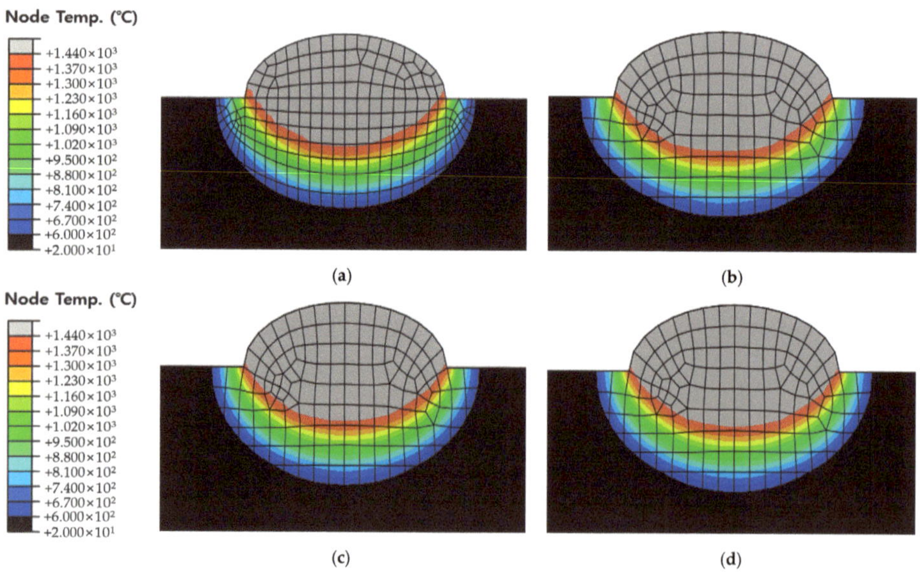

Figure 16. Comparison of temperature distribution of Case 3 and Original dimension model 1.6 s after the heat source has passed (**a**) Simplified model with fine meshes; (**b**) 18 mm position of original dimension model; (**c**) 86 mm position of original dimension model; (**d**) 186 mm position of original dimension model.

When visually comparing the temperature distribution of the simplified model and the original dimension model, it was confirmed that the sizes of the melted zone and HAZ area were similar until 1.6 s after the heat source passed. However, the measurement results for HAZ width and HAZ depth showed a slight difference. This is attributed to the original dimension model being a 3D model where heat transfer occurs in all directions.

Table 10 shows the HAZ width and HAZ depth compared to the simplified model when the heat source passes through the 18 mm, 86 mm, and 186 mm points of the original dimension model. In all cases, the size at 186 mm was significantly larger than at 18 mm. This is attributed to the heat generated by the heat source, indicating that heat transfer occurs effectively as the heat source moves. Subsequently, we compared the average measurements of the original dimension model based on the values from the simplified model as a reference. In Case 1, HAZ width and HAZ depth showed error rates of 11.01% and 12.75%, respectively. In Case 2, the error rates were 13.49% and 11.97%, and 12.63% and 9.53% in Case 3. All error rates were confirmed to be within 14%. Hence, it is considered advantageous to apply the simplified model in field scenarios where speed is crucial. Considering that the results from the simplified model are slightly higher than the actual experimental results (Table 9), we can conclude that the original dimension model provides heat transfer results with higher consistency. However, there is a need to develop a more sophisticated analysis model while ensuring high speed.

Table 10. HAZ size comparison between the original dimension and simplified models.

Value		Variables				
		Simplified Model (mm)	Original Dimension Model (mm)			Difference (%)
			18 mm	86 mm	186 mm	
HAZ Width	Case 1	13.83	12.25	12.28	12.39	11.01
	Case 2	14.70	12.60	12.71	12.84	13.49
	Case 3	16.89	14.62	14.70	14.95	12.63
HAZ Depth	Case 1	4.42	3.80	3.87	3.90	12.75
	Case 2	4.68	4.09	4.11	4.16	11.97
	Case 3	5.70	5.05	5.18	5.24	9.53

6. Conclusions

This sought to improve the slow interpretation speed, which hinders the procurement of arc welding heat sources, by applying optimization algorithms. More specifically, we aimed to improve the interpretation speed by devising simplified interpretation models. In addition, temperature constraints were added to compensate for the shortcomings of existing studies, and a new objective function was proposed to ensure consistency. The primary findings are summarized as follows:

(1) The heat transfer analysis results show that the optimal parameters of the Goldak model derived by the optimization algorithm satisfied all temperature constraints.

(2) The model applied in the previous study was simplified to speed up the analysis process, which increased the analysis speed by about 70%.

(3) A heat source model that melts the entire weld bead was derived through the temperature constraints of the weld point, and a consistent HAZ area was simulated through a new objective function.

(4) By comparing the HAZ width and HAZ depth of the simplified model with the actual weld cross-sections, the HAZ width and depth exhibited maximum differences of 4.81% and 14.73%, respectively.

(5) By comparing the heat transfer analysis results between the simplified model and the original dimension model, the HAZ width and depth showed maximum differences of 13.49% and 12.75%, respectively.

(6) This study applied a simplified model based on the HAZ size to rapidly derive optimal heat source parameters and identify the weld geometry through heat transfer analysis, which was considered to save time.

Author Contributions: Conceptualization, Y.K. and J.K.; Methodology, Y.K. and C.P.; Software, Y.K.; Validation, J.K., H.P., S.H., C.P. and G.P.; Investigation, H.P. and S.H.; Writing—original draft, Y.K.; Writing—review & editing, J.K., C.P. and G.P.; Supervision, J.K. and C.P. All authors have read and agreed to the published version of the manuscript.

Funding: This study has been conducted with the support of the Korea Institute of Industrial Technology as "Development of core technologies of AI based self-power generation and charging for next-generation mobility (KITECH EH-23-0013)" This study has been conducted with the support of the Ministry of Trade, Industry and Energy as "Development of welding and bonding technology for the 1st and 2nd barrier production process based on automation (P0018480)".

Institutional Review Board Statement: Not applicable.

Informed Consent Statement: Not applicable.

Data Availability Statement: The data presented in this study are available on request from the corresponding author.

Conflicts of Interest: The authors declare no conflict of interest.

References

1. Park, M.; Kim, J.; Pyo, C.; Son, J.; Kim, J. Research for the Optimal Flux-Cored Arc Welding Process of 9% Nickel Steel Using Multi Object Optimization with Solidification Crack Susceptibility. *Materials* **2021**, *14*, 1659. [CrossRef] [PubMed]
2. Park, D.-H.; Yi, M.-S.; Lee, J.-M. A Study on Welding-induced Temperature Distribution and Deformation of IMO Type C LNG Fuel Tank Support Structures. *J. Weld. Join.* **2022**, *40*, 118–132. [CrossRef]
3. Mohd Noor, C.W.; Noor, M.M.; Mamat, R. Biodiesel as alternative fuel for marine diesel engine applications: A review. *Renew. Sustain. Energy Rev.* **2018**, *94*, 127–142. [CrossRef]
4. Solakivi, T.; Laari, S.; Kiiski, T.; Töyli, J.; Ojala, L. How shipowners have adapted to sulphur regulations—Evidence from Finnish seaborne trade. *Case Stud. Transp. Policy* **2019**, *7*, 338–345. [CrossRef]
5. Vierth, I.; Karlsson, R.; Mellin, A. Effects of More Stringent Sulphur Requirements for Sea Transports. *Transp. Res. Procedia* **2015**, *8*, 125–135. [CrossRef]
6. Yang, Z.; Zhang, D.; Caglayan, O.; Jenkinson, I.; Bonsall, S.; Wang, J.; Huang, M.; Yan, X. Selection of techniques for reducing shipping NOx and SOx emissions. *Transp. Res. Part D Transp. Environ.* **2012**, *17*, 478–486. [CrossRef]
7. Gu, Y.; Wallace, S.W. Scrubber: A potentially overestimated compliance method for the Emission Control Areas: The importance of involving a ship's sailing pattern in the evaluation. *Transp. Res. Part D Transp. Environ.* **2017**, *55*, 51–66. [CrossRef]
8. Hwang, S.; Jeong, B.; Jung, K.; Kim, M.; Zhou, P. Life Cycle Assessment of LNG Fueled Vessel in Domestic Services. *J. Mar. Sci. Eng.* **2019**, *7*, 359. [CrossRef]
9. Burel, F.; Taccani, R.; Zuliani, N. Improving sustainability of maritime transport through utilization of Liquefied Natural Gas (LNG) for propulsion. *Energy* **2013**, *57*, 412–420. [CrossRef]
10. Lee, D.; Kim, K.H.; Choi, I.; Gil Lee, D. Pressure-resisting capability of the knot area of the primary barrier for a LNG containment system. *Ocean Eng.* **2015**, *95*, 128–133. [CrossRef]
11. Jo, Y.; Shin, K.; Hwang, S. Development of dynamic simulation model of LNG tank and its operational strategy. *Energy* **2021**, *223*, 120060. [CrossRef]
12. Wang, S. The Adoption of Liquefied Natural Gas as a Ship Fuel: A Systematic Review of Perspectives and Challenges. *Transp. Rev.* **2014**, *34*, 749–774. [CrossRef]
13. Kumar, S.; Kwon, H.-T.; Choi, K.-H.; Lim, W.; Cho, J.H.; Tak, K.; Moon, I. LNG: An eco-friendly cryogenic fuel for sustainable development. *Appl. Energy* **2011**, *88*, 4264–4273. [CrossRef]
14. Bernatik, A.; Senovsky, P.; Pitt, M. LNG as a potential alternative fuel—Safety and security of storage facilities. *J. Loss Prev. Process Ind.* **2011**, *24*, 19–24. [CrossRef]
15. Park, T.-U.; Jung, D.-H.; Park, J.-H.; Kim, J.-H.; Han, I.-W. Changes in the Mechanical Properties and Microstructure of High Manganese Steel by High Heat Input Welding and General Welding Processes. *J. Weld. Join.* **2022**, *40*, 33–39. [CrossRef]
16. Park, T.-U.; Jeong, Y.-C.; Im, H.-D.; Choi, C.; Kil, W. Development and Evaluation of Stainless Steel-Base Flux-Cored Wires for 9% Nickel Steel. *J. Weld. Join.* **2022**, *40*, 367–376. [CrossRef]
17. Pyo, C.-M. Experimental Study According to the Type of Shielding Gas using Fiber Laser Welding of 9% Nickel Steel. *J. Korean Soc. Mech. Technol.* **2021**, *23*, 702–707. [CrossRef]
18. Park, J.-H.; Kim, Y.-H.; Baek, H.-J.; Cho, S.-M. A study on process development of super-TIG welding for 9% nickel steel with Alloy 625. *J. Manuf. Process.* **2019**, *40*, 140–148. [CrossRef]
19. Xu, T.; Shi, Y.; Jiang, Z.; Wu, L.; Ma, Y.; Wang, Z. Improvement of cryogenic toughness for 9% Ni steel keyhole TIG butt-welded joints with a Ni interlayer. *Mater. Sci. Eng. A* **2022**, *835*, 142661. [CrossRef]
20. Gook, S.; Krieger, S.; Gumenyuk, A.; El-Batahgy, A.; Rethmeier, M. Notch impact toughness of laser beam welded thick sheets of cryogenic nickel alloyed steel X8Ni9. *Procedia CIRP* **2020**, *94*, 627–631. [CrossRef]

21. Kim, J. A Study on FCAW Butt Welds of 9% Nickel Steel Materials for Vessel Cryogenic Tank Part II: Mechanical Properties Study of Welding Bead. *J. Korean Soc. Mech. Technol.* **2021**, *23*, 426–431. [CrossRef]
22. Mu, W.; Li, Y.; Cai, Y.; Wang, M. Cryogenic fracture toughness of 9%Ni steel flux cored arc welds. *J. Mater. Process. Technol.* **2018**, *252*, 804–812. [CrossRef]
23. Park, M.; Kim, J.; Pyo, C.; Kim, J.; Chun, K. A Study on the Weldment Hardening Discrimination Procedure and Improvement of Flux Cored Arc Welding Process of ASTM A553-1 (9% Nickel Steel) Material Using Bead Geometry Distribution. *Metals* **2021**, *11*, 1282. [CrossRef]
24. Moshtaghi, M.; Loder, B.; Safyari, M.; Willidal, T.; Hojo, T.; Mori, G. Hydrogen trapping and desorption affected by ferrite grain boundary types in shielded metal and flux-cored arc weldments with Ni addition. *Int. J. Hydrogen Energy* **2022**, *47*, 20676–20683. [CrossRef]
25. Kim, Y.-K.; Kim, J.-H. Welding Residual Stress and Strength of Thick 9% Nickel Steel Plate. *J. Korean Soc. Power Syst. Eng.* **2014**, *18*, 85–90. [CrossRef]
26. Manurung, Y.H.P.; Lidam, R.N.; Rahim, M.R.; Zakaria, M.Y.; Redza, M.R.; Sulaiman, M.S.; Tham, G.; Abas, S.K. Welding distortion analysis of multipass joint combination with different sequences using 3D FEM and experiment. *Int. J. Press. Vessel. Pip.* **2013**, *111–112*, 89–98. [CrossRef]
27. Deng, D.; Murakawa, H. FEM prediction of buckling distortion induced by welding in thin plate panel structures. *Comput. Mater. Sci.* **2008**, *43*, 591–607. [CrossRef]
28. García-García, V.; Mejía, I.; Reyes-Calderón, F. Experimental and FEM study of Ti-containing TWIP steel weldability. *J. Mater. Process. Technol.* **2018**, *261*, 107–122. [CrossRef]
29. Pyo, C.-M. Research for Estimation of Heat Source Model Parameters with Simplified FE Model & Global Optimization Algorithm Part I. Application of Adaptive Simulated Annealing. *J. Korean Soc. Mech. Technol.* **2020**, *22*, 484–491. [CrossRef]
30. Park, M.; Kim, J.; Pyo, C. A Study on Heat Input Control and a Quality Evaluation Algorithm to Prevent Toughness Deterioration of the Heat-Affected Zone in the Fiber Laser Welding Process of ASTM A553-1 (9% Nickel Steel) Material. *Metals* **2022**, *12*, 1195. [CrossRef]
31. Obeid, O.; Leslie, A.J.; Olabi, A.G. Influence of girth welding material on thermal and residual stress fields in welded lined pipes. *Int. J. Press. Vessel. Pip.* **2022**, *200*, 104777. [CrossRef]
32. Attar, M.A.; Ghoreishi, M.; Beiranvand, Z.M. Prediction of weld geometry, temperature contour and strain distribution in disk laser welding of dissimilar joining between copper & 304 stainless steel. *Optik* **2020**, *219*, 165288. [CrossRef]
33. Pankaj, P.; Sawarkar, P.S.; Tiwari, A.; Biswas, P.; Pal, S. Three dimensional FE thermal analysis for friction stir welding of low carbon steel. *Mater. Today Proc.* **2020**, *41*, 902–907. [CrossRef]
34. Hammad, A.; Churiaque, C.; Sánchez-Amaya, J.M.; Abdel-Nasser, Y. Experimental and numerical investigation of hybrid laser arc welding process and the influence of welding sequence on the manufacture of stiffened flat panels. *J. Manuf. Process.* **2020**, *61*, 527–538. [CrossRef]
35. Ling, Y.; Ni, J.; Antonissen, J.; Ben Hamouda, H.; Voorde, J.V.; Wahab, M.A. Numerical prediction of microstructure and hardness for low carbon steel wire Arc additive manufacturing components. *Simul. Model. Pract. Theory* **2023**, *122*, 102664. [CrossRef]
36. Hu, L.; Li, X.; Luo, W.; Li, S.; Deng, D. Residual stress and deformation in UHS quenched steel butt-welded joint. *Int. J. Mech. Sci.* **2023**, *245*, 108099. [CrossRef]
37. Li, X.; Hu, L.; Deng, D. Influence of contact behavior on welding distortion and residual stress in a thin-plate butt-welded joint performed by partial-length welding. *Thin-Walled Struct.* **2022**, *176*, 109302. [CrossRef]
38. Luo, W.; Guo, X.; Dai, J.; Rao, T. Hull optimization of an underwater vehicle based on dynamic surrogate model. *Ocean Eng.* **2021**, *230*, 109050. [CrossRef]
39. Farias, R.; Teixeira, P.; Vilarinho, L. Variable profile heat source models for numerical simulations of arc welding processes. *Int. J. Therm. Sci.* **2022**, *179*, 107593. [CrossRef]
40. Farias, R.; Teixeira, P.; Vilarinho, L. An efficient computational approach for heat source optimization in numerical simulations of arc welding processes. *J. Constr. Steel Res.* **2020**, *176*, 106382. [CrossRef]
41. Pyo, C.; Jeong, S.-M.; Kim, J.; Park, M.; Shin, J.; Kim, Y.; Son, J.; Kim, J.-H.; Kim, M.-H. A Study on the Enhanced Process of Elaborate Heat Source Model Parameters for Flux Core Arc Welding of 9% Nickel Steel for Cryogenic Storage Tank. *J. Mar. Sci. Eng.* **2022**, *10*, 1810. [CrossRef]
42. Wang, H.; Hou, Y.; Xiong, Y. Research on multi-interval coupling optimization of vessel speed for energy efficiency. *Ocean Eng.* **2022**, *257*, 111559. [CrossRef]
43. Pyo, C.-M. A Study for Heat Source Model for Fiber Laser Welding with Global Optimization Algorithm Part I. Verification of Multi Layered Heat Source Model. *J. Korean Soc. Mech. Technol.* **2021**, *23*, 137–144. [CrossRef]
44. Hahn, Y.; Cofer, J. Design Study of Dovetail Geometries of Turbine Blades Using Abaqus and Isight; Volume 7: Structures and Dynamics, Parts A and B. In Proceedings of the ASME Turbo Expo 2012: Turbine Technical Conference and Exposition, Copenhagen, Denmark, 11–15 June 2012; pp. 11–20.
45. He, K.; Tole, K.; Ni, F.; Yuan, Y.; Liao, L. Adaptive large neighborhood search for solving the circle bin packing problem. *Comput. Oper. Res.* **2020**, *127*, 105140. [CrossRef]
46. Hifi, M.; Paschos, V.T.; Zissimopoulos, V. A simulated annealing approach for the circular cutting problem. *Eur. J. Oper. Res.* **2004**, *159*, 430–448. [CrossRef]

47. Geng, X.; Chen, Z.; Yang, W.; Shi, D.; Zhao, K. Solving the traveling salesman problem based on an adaptive simulated annealing algorithm with greedy search. *Appl. Soft Comput.* **2011**, *11*, 3680–3689. [CrossRef]
48. Kirkpatrick, S.; Gelatt, C.D., Jr.; Vecchi, M.P. Optimization by Simulated Annealing. *Science* **1983**, *220*, 671–680. [CrossRef]
49. Jeong, H.-M. *Principles of Intelligent Information Systems for the 21st Century*; 21cbook: Paju City, Republic of Korea, 1997; pp. 354–375, ISBN 978-898-700-116-6.
50. Navid, M.; Soheil, G.; Behzad, A.; Izman, S.; Hamidreza, G.; Norizah, R.; Shukur, H.; Amran, A.; Sehun, R. A novel systematic numerical approach on determination of heat source parameters in welding process. *J. Mater. Res. Technol.* **2022**, *18*, 4427–4444. [CrossRef]

materials

MDPI

Article

Microstructure and Mechanical Properties of Steel and Ni-Based Superalloy Joints for Rotors of High-Speed Electric Motors

Eero Scherman [1,2,*], Eerik Sikanen [1], Hemantha Kumar Yeddu [1], Mohsen Amraei [1,3] and Jussi Sopanen [1]

1 Mechanical Engineering, LUT School of Energy Systems, LUT University, P.O. Box 20, 53851 Lappeenranta, Finland
2 Department of Technology, LAB University of Applied Sciences, Yliopistonkatu 36, 53850 Lappeenranta, Finland
3 Department of Mechanical and Materials Engineering, University of Turku, 20520 Turku, Finland
* Correspondence: eero.scherman@lab.fi; Tel.: +358-406-747-132

Abstract: High-speed electric motors, e.g., axially laminated anisotropic synchronous reluctance motors (ALA-SynRM), use a solid rotor manufactured by joining alternating layers of magnetic and non-magnetic metallic sheets. The strength of the dissimilar metallic joints is critical for the rotor's ability to withstand the operating conditions of the high-speed electrical machine. In this work, various dissimilar metallic joint configurations that can be used in high-speed ALA-SynRM rotors are studied by analyzing the shear strength, microstructure, hardness, and composition of the joints. Metallic joints of structural steels and Inconel® alloys fabricated by vacuum brazing and hot isostatic pressing (HIP) are studied. Finite element analysis (FEA) was performed to calculate the maximum shear stress of the joints that were subjected to various high speed operating conditions. The shear strength of the test specimens was measured and compared with FEA results. The microstructure and chemical composition of the joints were studied by using optical microscopy, scanning electron microscopy (SEM) and energy dispersive spectroscopy (EDS) on SEM. The results show that the hot isostatic pressed S1100MC-IN718 joint achieved the highest ultimate shear strength (233.3 MPa) followed by vacuum brazed S355MC-IN600 joint (230.1 MPa) and HIP S355-IN718 (203.5 MPa), thereby showing that vacuum brazing and HIP can be viable manufacturing methods to fabricate a high-speed ALA-SynRM rotor.

Keywords: iron alloys; nickel alloys; shear strength; microstructure; grains and interfaces; finite element analysis

Citation: Scherman, E.; Sikanen, E.; Yeddu, H.K.; Amraei, M.; Sopanen, J. Microstructure and Mechanical Properties of Steel and Ni-Based Superalloy Joints for Rotors of High-Speed Electric Motors. *Materials* **2022**, *15*, 6906. https://doi.org/10.3390/ma15196906

Academic Editor: Antonio Riveiro

Received: 29 August 2022
Accepted: 29 September 2022
Published: 5 October 2022

1. Introduction

High-speed (HS) electrical machine (EM) technologies offer a viable solution for the needs of future electromechanical energy conversion systems. A key benefit of high-speed technology is its higher power density, which decreases the size of high-speed systems and reduces the need of raw materials. Moreover, high-speed machines exhibit greater efficiency than electrical machines with lower rotational speeds [1]. Typical applications for HS EM are, for example, gas turbines, flywheel energy storage systems, high-speed spindle applications, turbomolecular pumps, gas compressor applications, as well as industrial air compressors and air blowers [2]. An important aspect of electromagnetic design and manufacturing of such high-speed technology is rotor dynamics, which influences robustness and efficiency, and consequently the economics of the rotating electrical machine. High-speed rotating electrical machines are electrical machines with rotor surface speed greater than 100 m/s [3] and high rotational speed with respect to the power of the electrical machine [2]. The maximum operation speed of a HS-EM is governed by the mechanical robustness of its rotor at high speeds [4], thermal losses in the rotor and in the stator occurring at high switching frequencies [5], and gas friction losses at rotor surface speed over 300 m/s [6].

Typically, high-speed electrical machines can be classified by machine type and rotor structure into induction machines (IM), permanent magnet synchronous machines (PMSM) and switched reluctance machines (SRM) [7]. Synchronous homopolar machines [8] and synchronous reluctance motors (SynRM) [9] are also applicable to be used in high-speed applications.

Each of the high-speed electrical machine types listed above has a unique rotor structure which makes the machine particularly suitable for specific application. The strengths and weaknesses of the various high-speed electrical machine types are described in [4]. As a summary, it can be concluded that induction machines are generally suited for high-speed applications due to the mechanical robustness of the IM rotor. However, the efficiency of the IM rotor is decreased by rotor joule losses. Switched reluctance machine rotors are simple and robust constructions, but the requirement of a large airgap reduces their efficiency. Permanent magnet synchronous machines have high power density and efficiency; however, the mechanical robustness of the rotor, which is affected by the magnet encapsulation, may be a limiting factor in high-speed applications.

To overcome some of the limitations of conventional high-speed rotors noted above, an axially laminated anisotropic synchronous reluctance motor (ALA-SynRM) was presented in [7]. The ALA-SynRM rotor is a solid rotor structure made from multiple alternating magnetic and non-magnetic metal sheets that provide the necessary flux paths for the electromagnetic field. A key benefit of the ALA-SynRM rotor is its simplicity. Unlike PMSM rotors, the ALA-SynRM rotor does not require rare earth permanent magnets. Moreover, there is no need for the complicated squirrel-cage structure found in IM rotors, which can limit the maximum rotational speed of the rotor by decreasing the rigidity of the rotor [10].

Typical methods for bonding dissimilar metals sheets includes brazing and high and low temperature solid-state welding processes, such as explosion welding and hot isostatic pressing (HIP), all of which can be used to manufacture ALA-SynRM rotors [11]. The brazing process is described as a process where the parent materials to be bonded are heated to a temperature that is lower than the melting temperature of the parent materials but higher than the melting temperature of the braze alloy [12]. As the braze alloy melts in the joint, capillary forces and the wetting action cause the braze alloy to flow throughout the joint area and, on cooling, to create a permanent bond between the workpieces. In the explosion welding process, an explosive charge is used to accelerate the parts to be bonded to high speeds of up to several hundreds of meters per second. The parts collide with each other and solid bonding of the two metal pieces occurs at the point of impact [13]. The HIP process is used as a solid-state diffusion welding process to bond dissimilar materials and to sinter powder metal components with fully isotropic material parameters [14]. The joint is formed by deforming and diffusing the joint area using high temperature and pressure [15].

Recently, the electromagnetic properties of ALA-SynRM rotors manufactured by mechanically bonding the laminated rotor structure have been studied in [16–18] and all aforementioned researchers highlight similar or higher rotor efficiency compared to other rotor structures. These studies, however, consider only ALA-SynRM rotors manufactured by mechanically joining, e.g., with screws, the alternating magnetic and non-magnetic metallic sheets. Using external elements to construct the ALA-SynRM rotor can lead to a complicated construction process in terms of manufacturing and may compromise the structural integrity of the rotor in high-speed applications [19]. The utilization of metal-to-metal bonding methods, i.e., brazing, explosion welding and hot isostatic pressing, to manufacture ALA-SynRM rotor structures has not been widely studied. While the feasibility of a brazed ALA-SynRM rotor was studied in [20], it was primarily focused on describing the electromagnetic performance of the brazed ALA-SynRM rotor with little consideration given to joint strength and structure.

The detailed analysis of EM rotors structural performance and its applicability in high-speed operations in [21] highlighted the importance of effective bonding of the dissimilar materials. Accordingly, this article investigates dissimilar metal joints suitable for use in construction of the rotor of an ALA-SynRM. The mechanical properties, such as shear strength, and microstructural aspects of the joints made from several different materials are studied. The paper begins by giving an overview of the structure of an ALA-SynRM rotor. The subsequent section introduces the experimental methods used to determine the shear strength of the joints followed by finite element analysis of the stresses that the rotor is subjected to during its operation. Finally, the shear strength measurements and the corresponding microstructural aspects are discussed. The paper concludes by reprising key findings from the work.

2. Materials and Methods

During operation of an electrical machine, the rotor is subjected to mechanical loads caused by the rotational motion of the rotor and electromagnetic forces created by the electrical machine. Additionally, the rotor may be subjected to thermal loads due to varying operational temperature. The mechanical robustness of the rotor is often the limiting factor for high rotational speeds.

In the case of an ALA-SynRM rotor, the rotor is constructed from multiple magnetic and non-magnetic sheets bonded to each other. A two-pole ALA-SynRM rotor cross section is shown in Figure 1. In the ALA-SynRM rotor, the magnetic sheets act as a flux path and the non-magnetic sheets act as insulators that guide the electromagnetic flux [22]. The strength of the joint combining the sheets depends on the strength of the parent material and the joint integrity which are studied using experiments and simulations as explained in the following sections.

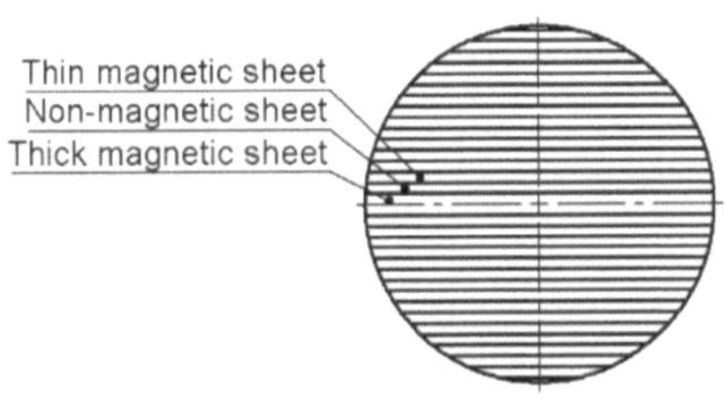

Figure 1. Cross section of ALA-SynRM rotor (according to [7]).

2.1. Experiments

2.1.1. Materials

This study focuses on the strength and structure of the following three bimetallic joint combinations: (i) vacuum brazed joint of structural steel S355MCD and non-precipitation hardened Inconel® 600, (ii) hot isostatically pressed structural steel S355MC and precipitation hardened Inconel® 718, and (iii) ultra-high strength structural steel S1100MC and precipitation hardened Inconel® 718. Material pairs for this research were selected based on existing literature [23], as well as the aim of studying the possibilities of using ultra-high strength structural steel in the construction of an ALA-SynRM rotor. The chemical composition of the parent materials and braze alloy for vacuum brazing of this study are presented in Appendix A Table A1.

Table 1 presents the thickness, yield strength, ultimate tensile strength, elongation, hardness, and delivery state of the materials and braze alloy studied at a temperature of 20 °C in the delivery state as indicated in the material certificates. For S1100, Inconel® 600, Inconel® 718, and braze alloy CW021A, yield strength is 0.2% proof stress (Rp0.2).

Table 1. Mechanical properties of the parent materials based on material certificates.

Material	Thickness (mm)	Yield Strength (MPa)	Tensile Strength (MPa)	Elongation (%)	Hardness	Delivery State
S355MC	3.00	402	487	34 (A45)	155 (HV3/10) [1]	Thermomechanically formed
S355MCD	6.00	421	490	34 (A50)	155 (HV3/10) [1]	Thermomechanically formed
S1100MC	4.00	1166	1363	9.0 (A5)	453 (HV3/10) [1]	Thermomechanically formed
S1100MC	6.00	1144	1374	9.0 (A5)	453 (HV3/10) [1]	Thermomechanically formed
Inconel® 600	3.00	327	691	48.1 (A50)	86.2 (HRBW)	Soft annealed
Inconel® 718	3.17	419	878	53	92 (HRBW)	Annealed
Inconel® 718	3.71	450.2	903	50.4	95.3 (HRB)	Annealed
CW021A	0.100	98	247	36 (A50)	-	Soft annealed

[1] measured value (Struers Durascan, HV3/10, Denmark).

2.1.2. Bonding Processes

Figure 2 shows the shape of the billet used to construct the ALA-SynRM rotor and the shape of the test specimens extracted from the rotor billet.

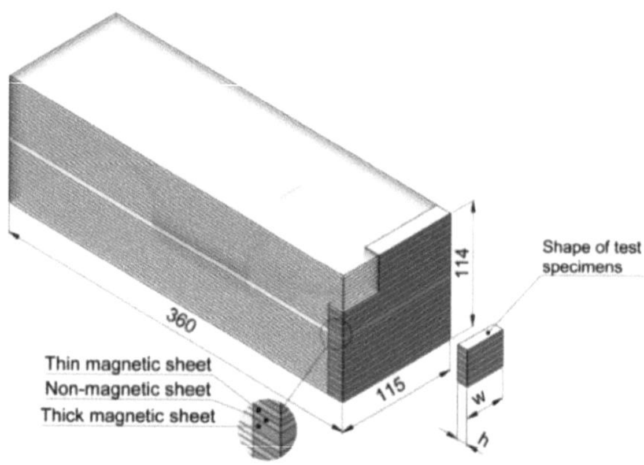

Figure 2. Structure of the billet for ALA-SynRM rotors and test specimens (dimensions in mm, not to scale).

Table 2 describes the bonding methods and test specimens used in this study. Test specimens to study HIP bonding were manufactured from the excess material of the billets used to manufacture the ALA-SynRM rotors (Figure 2). To study the vacuum brazed joint, a separate test billet was manufactured after the manufacturing of the rotor billet, since the rotor billet did not provide enough test specimens to study the joint. The vacuum brazed test specimen contains only thick magnetic sheets to provide a more favorable test procedure in practice.

Table 2. Test specimens and their bonding methods and process parameters.

Test specimen	Material	Thickness (mm)	Bonding Method	Temperature (°C) and Dwell Time (h)	Pressure at Maximum Temperature (MPa)
A	S355MCD Inconel® 600	6.00 3.00	Vacuum brazing	1100/0.5	0.0001
B	S1100MC Inconel® 718	4.00, 6.00 3.17	Hot isostatic pressing	1150/5	101
C	S355MC Inconel® 718	3.00, 6.00 3.17	Hot isostatic pressing	1150/4	101
D	S355MC	6.00	-	-	-
E	S1100MC	4.00	-	1100/5	-
F	Inconel® 718	3.71	-	1100/5	-
G	Inconel® 600	3.00	-	1100/1	-
H	S355MCD	6.00	-	1100/1	-

Control specimens D, E, F, G and H were included in the experiment. Specimen D was used to validate the shear strength test setup by measuring the shear strength of a parent material with known strength properties. Control specimens E, F, G and H were used to determine the tensile strength of the parent materials after heat treatment similar to the test specimens used in the bonding processes.

Manufacturing of the rotor and test specimen billets began by cutting the metal plates to suitable shapes with a fiber laser. After preparation, the cut plates were sent to companies specializing in the vacuum brazing and hot isostatic bonding processes. For test specimen A, the rate of heating in the brazing process was 314 °C/h before 30 min dwell time at a brazing temperature of 1100 °C. In the brazing atmosphere, the pressure was less than 0.0001 MPa during the brazing process. The cooling cycle lasted 2 h, during which the test piece cooled down to 950 °C in the first hour and to room temperature in the second hour. A copper braze alloy with thickness of 100 µm was utilized in the vacuum brazing while the strength and ductility of the braze alloy material was considered to be suitable on this application.

Authors in [23] suggested that HIP should be used when bonding Inconel® 718 because vacuum brazing of Inconel® 718 and S355 structural steel was experimented with copper braze alloy and results were unsatisfactory. Thus, the material pairs including Inconel® 718 in this research are bonded by HIP. In the HIP process, the rate of heating was 164 °C/h and the rate of pressure increase was 14.4 MPa/h. The rate of cooling was 192 °C/h and the rate of pressure decrease was 16.3 MPa/h. The HIP process was performed in an argon-protected atmosphere. After successful bonding, all test specimens were visually inspected before machining to a shape suitable for the test setup.

The control specimens E and F were exposed to a high temperature of 1100 °C for a long time (5 h) to simulate the effect of long-term exposure to high temperature simulating HIP process. The control specimens G and H were exposed to a similar temperature as in the vacuum brazing for one hour to simulate the temperature cycle of vacuum brazing. All control specimens were encapsulated in individual argon protected atmospheres during exposure to high temperature.

2.1.3. Experimental Testing

When assessing the strength of a joint, either the normal or the shear strength can be studied. Due to the structure of the ALA-SynRM rotor, the latter is the focus of this research. Shear stress along a lap joint can be evaluated by using test specimen geometry as demonstrated in [24] or directly induce shear stress on the test specimen with a dedicated test setup as in [25]. In this study, a test setup adopted from [25] was created and Figure 3 illustrates the test setup and the principle of operation.

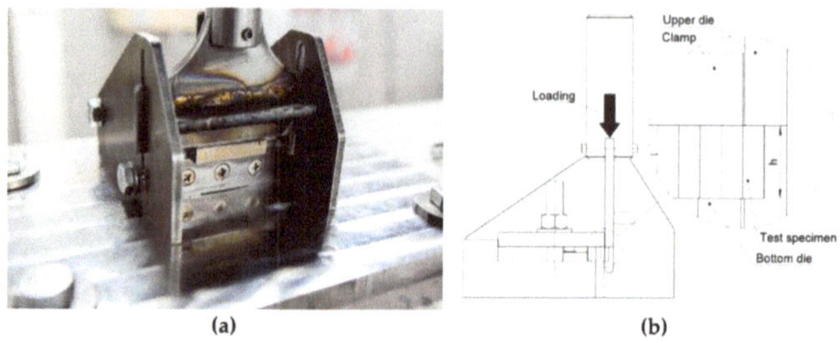

Figure 3. Test setup for the shear strength (**a**) experimental setup; (**b**) schematic drawing.

In the study, the test specimen was carefully placed such that the joint was located between the upper and lower die. In each test, the structural steel was located under the upper die and, respectively, Inconel® was always on top of the bottom die.

The shear strength of the joints was determined using a universal testing machine (Matertest Oy FMT-250, Finland) equipped with a servo-hydraulic cylinder with a maximum force capacity of 250 kN. Test specimens were subject to a load which was gradually increased up to the breaking point with the strain rate indicated in Table 3. The universal testing machine was equipped with a calibrated load cell (Interface Inc. 1020AF-125KN-B, USA), which allowed the force exerted to be recorded during the test. The servo-hydraulic cylinder was equipped with a position transducer (Curtiss-Wright Corp., Penny & Giles VRVT100, United Kingdom), which allowed the displacement of the die during the tests to be recorded. The universal testing machine was also used to determine the ultimate tensile strength of the control specimens E, F, G and H.

Table 3. Test and test specimen details.

Test Specimen	Test Type	Material	Bonding Method	Width (mm)	Height (mm)	Gauge Length (mm)	Strain Rate (s^{-1})
A	Shear	S355MCD Inconel® 600	Vacuum brazing	50	10	-	0.001
B	Shear	S1100MC Inconel® 718	Hot isostatic pressing	36	5	-	0.002
C	Shear	S355MC Inconel® 718	Hot isostatic pressing	50	6	-	0.002
D	Shear	S355MCD	-	50	6	-	0.002
E	Tensile	S1100MC	-	30	4	50	0.0002
F	Tensile	Inconel® 718	-	30	3.71	50	0.0002
G	Tensile	Inconel® 600	-	30	3	50	0.0002
H	Tensile	S355MCD	-	30	6	50	0.0002

Metallographic samples were prepared and etched using 4% Nital for 10 s for structural steel samples and 10% oxalic acid with 30 V/2 A DC-current for Inconel samples. The microstructure of the specimens was examined with a stereo microscope (Meiji Techno IM7530, Japan). The microhardness of the specimens was measured with a micro hardness tester (Struers Durascan, HV3/10, Denmark). A scanning electron microscope (Hitachi SU3500, Japan) was used to study the composition of the joint area in greater detail.

2.2. Finite Element Simulations

The design parameters for the rotor under study are as follows: maximum rotational speed of 24,000 rpm and nominal power of 12 kW. The laminate layer thicknesses and configuration are shown in Figure 4. The materials initially considered for the finite element (FE) study of the laminated rotor are S355 steel for the magnetic material and Inconel® 718 for the non-magnetic material, due to the following reasons. In addition to the excellent electromagnetic performance, this material pair has very similar structural properties, which minimizes the difference in thermal growth under thermal load and, in general, the overall strain difference between the laminated sheets. In addition, the availability and price of the materials are favorable. Finite element analysis (FEA) was performed also for other material pairs shown in Table 3. The material properties used in FEA are shown in Table 4. The software used for finite element analysis was Ansys Workbench 2021 R2.

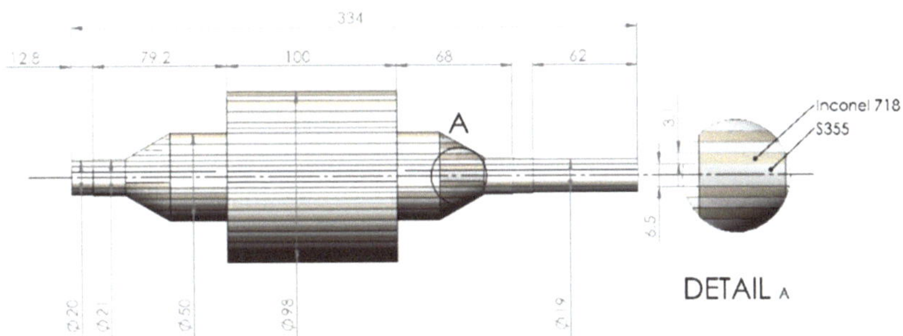

Figure 4. Main dimensions and laminate layer configuration of the laminated rotor. (Dimensions in millimeters, not to scale.)

Table 4. Material properties used in finite element analysis of the laminated rotor [26–29].

Material	Density (kg/m^3)	Elastic Modulus (GPa)	Poisson's Ratio (-)	Thermal Expansion (1/°C)
S355	7800	210	0.30	12.0×10^{-6}
Inconel® 718	8190	200	0.29	12.5×10^{-6}
S1100	7850	210	0.30	11.0×10^{-6}
Inconel® 600	8470	214	0.32	10.4×10^{-6}

Based on the CAD geometry depicted in Figure 4, a finite element (FE) model was generated to be able to solve the stresses the structure experiences under certain loading conditions. The rotational velocity was set to nominal speed 24,000 rpm, and the estimated thermal condition was included in the model by applying constant 300 °C increase from the initial condition. To maintain the size of the numerical problem feasible for solution, bonding material layers and the thermal loading cycle during the bonding process were excluded from the analysis. Instead, the laminated layers were directly connected using fixed constraints to simulate perfect bonding of laminated layers. Mesh density was adjusted to have a minimum of two element layers in the direction of thickness in each laminate, resulting in a total of 1,103,000 nodes in a mixed hexahedron dominant mesh having a total of 353,000 quadratic elements. This particular mesh density was utilized to optimize the computational time with the hardware resources available and it was deemed appropriate for the purposes of comparative analysis of the joints. In all the finite element analyses, the model geometry and the mesh density were kept unchanged, to be able to focus the analysis only on the behavior of different materials under the stresses.

3. Results

3.1. Experiments

3.1.1. Shear Strength and Hardness

The data gathered during the shear strength tests was analyzed and used to assess the shear behavior of the bimetallic joints. The engineering value of the shear stress was determined by dividing the force exerted on the test piece by the cross-section area of the test piece. Due to the nature of the test setup, only the displacement of the upper die (Figure 3b) could be recorded from the data provided by the position transducer in the servo-hydraulic cylinder. To obtain shear stress versus the displacement of the die, the displacement was normalized with the specimen thickness similarly as [30].

Figure 5 presents the shear stress-normalized displacement for the test specimens, and Tables 5 and 6 show the results of the shear strength of the joints and the ultimate tensile strength of the control specimens, respectively.

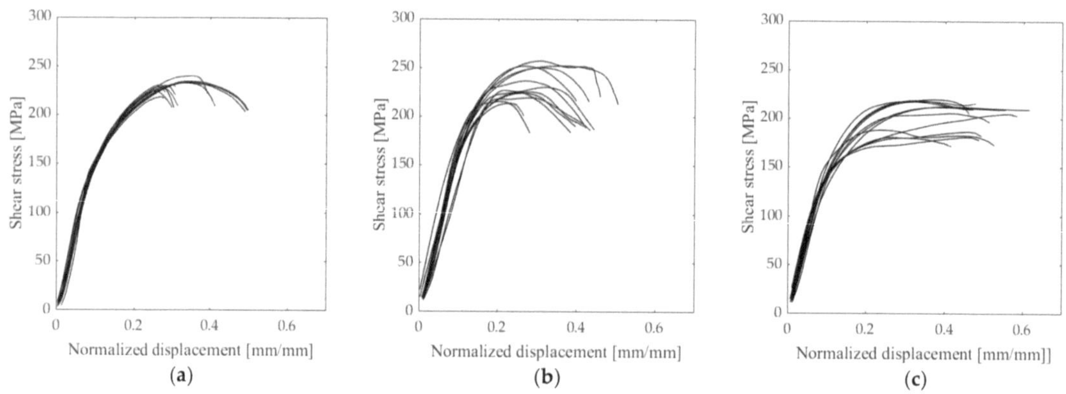

Figure 5. Shear stress-normalized displacement of test specimens: (**a**) S355-IN600 vacuum brazed; (**b**) S1100-IN718 HIP; and (**c**) S355-IN718 HIP.

Table 5. Shear strength of the test specimens (n = number of samples).

Test Specimen	Material	Bonding Method	n	Mean Ultimate Shear Strength (Mpa)	Standard Deviation of the Ultimate Shear Strength (Mpa)
A	S355MCD Inconel® 600	Vacuum brazing	12	230.1	5.5
B	S1100MC Inconel® 718	Hot isostatic pressing	12	233.3	15.7
C	S355MC Inconel® 718	Hot isostatic pressing	12	203.5	15.5
D	S355MCD	-	10	300.3	13.3

Table 6. Tensile strength of the control specimens (n = number of samples).

Test Specimen	Material	n	Mean Ultimate Tensile Strength (Mpa)	Standard Deviation of the Ultimate Tensile Strength (Mpa)
E	S1100MC	5	755.3	20.2
F	Inconel® 718	5	1317.5	17.2
G	Inconel® 600	5	740.9	30.5
H	S355MCD	5	352.7	27.1

According to the shear strength tests, the shear strength of the vacuum brazed S355-Inconel® 600 joint and S1100-Inconel® 718 HIP joint is almost equal regardless of the different bonding methods and different parent materials. It is noteworthy that the joint configuration S355-Inconel® 718 HIP produced the lowest shear strength of the study.

Test specimen D was used to determine the shear strength of the parent material S355MCD as a reference for the test setup. The results showed an ultimate shear strength of 300.3 Mpa (n = 10, SD = 13.3 Mpa). The material certificate for S355MCD showed an ultimate tensile strength of 490 Mpa (Table 1). Considering the von Mises yield criterion, the measured ultimate shear strength of the parent material is slightly higher than the theoretical value ($490\ MPa \times \frac{1}{\sqrt{3}} \approx 285\ MPa$).

Figure 6a–c describe the macro structure and micro hardness profile (HV3/10) of the joint in the vacuum brazing and HIP test specimens. The microhardness profile of the S355MCD-IN600 test specimen does not exhibit major changes in the hardness of the materials compared to the hardness at delivery state. Only minor loss of hardness compared to the delivery state can be noted in the structural steel S355MCD. However, the hardness profile of the S1100MC-IN718 test specimen displays a significant change in the hardness compared to the delivery state of the parent materials. The hardness of Inconel® 718 shows high increase in hardness compared to the parent material in the delivery state and the hardness of structural steel S1100MC is less than half of the original hardness as given in Table 1. The joint configuration of structural steel S355MC-IN718 exhibits similar behavior to the S1100MC-IN718 test specimen.

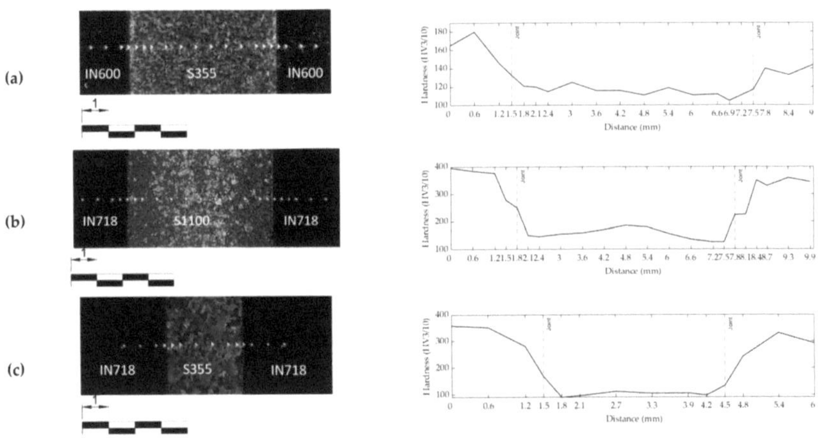

Figure 6. Macroscopic structure of the joints and microhardness (HV3/10): (**a**) S355-IN600 vacuum brazed; (**b**) S1100-IN718 HIP; and (**c**) S355-IN718 HIP.

3.1.2. Microstructure

Figure 7 shows the microstructures of the parent materials in the delivery state and after bonding. Additionally, average grain size ($\overline{P_L}$) of each parent material in the delivery state and after bonding was determined using a mean linear intercept procedure following standard SFS-EN ISO 643:2020. Due to the complex microstructure of structural steel S1100MC, the average grain size in the delivery state could not be determined by using the mean linear intercept method.

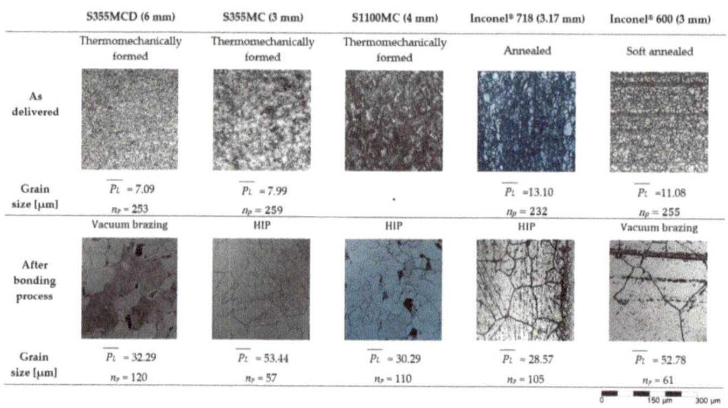

Figure 7. Microstructures of the test specimens (n_p = number of intersections).

All the test specimens show considerable change in the microstructures before and after the joining processes. For structural steel S355MC, the fine-grained ferrite–pearlite microstructure has changed to a coarse-grained ferrite–pearlite microstructure following the vacuum brazing and hot isostatic bonding processes. The microstructure of the as-delivered thermomechanically formed S1100MC steel is a mixture of bainite and martensite with retained austenite islands scattered along grain boundaries [31], but after the HIP bonding, the microstructure is coarse-grained ferrite–pearlite.

The microstructure of Inconel® 718 as-delivered state is solid solution austenite (γ) with minor phases γ', γ'', and δ [32]. After the HIP bonding, the microstructure has a small grain size in the joint area and relatively large grain size farther away from the joint area. Twinning is also apparent in the microstructure, which is identified by the straight lines across the grains. Inconel® 600 has an austenitic microstructure with titanium nitride and titanium carbide precipitates in the soft annealed state [28]. After the vacuum brazing, the microstructure of the Inconel® 600 has a very large grain size, similar to the Inconel® 718 test specimen.

Scanning electron microscopy (SEM) was used to gain a better understanding of the chemical composition of the joint; see Figures 8–10. The chemical composition was determined by energy dispersive spectroscopy (EDS) using secondary electron (SE) detector. The amounts of different alloying elements, in weight percentage, in each joint is presented with a color legend.

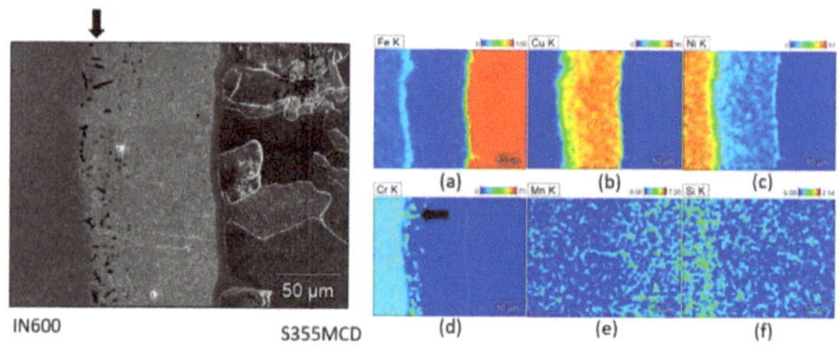

Figure 8. Micrographs of the structure and composition in weight % of the S355MCD-IN600 joint area: (**a**) iron (**b**) copper (**c**) nickel (**d**) chromium (**e**) manganese (**f**) silicon.

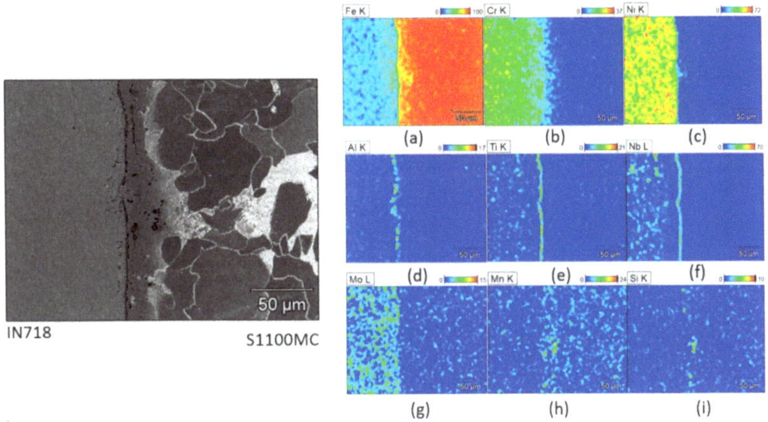

Figure 9. Micrographs of the structure and composition in weight % of the S1100MC-IN718 joint area: (**a**) iron (**b**) chromium (**c**) nickel (**d**) aluminium (**e**) titanium (**f**) niobium (**g**) molybdenum (**h**) manganese (**i**) silicon.

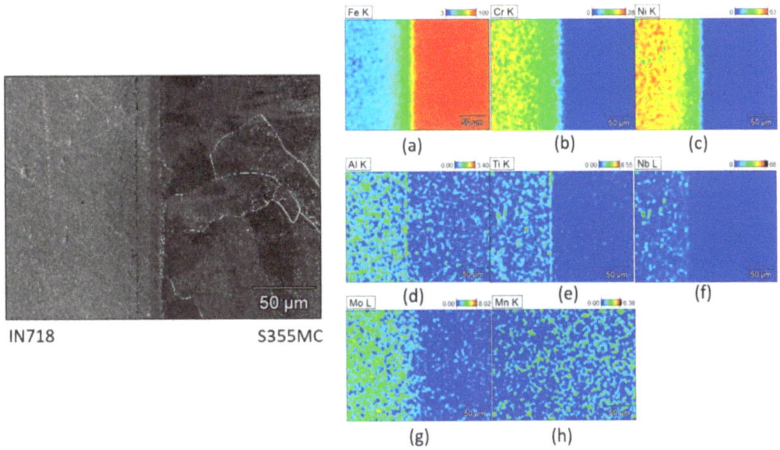

Figure 10. Micrographs of the structure and composition in weight % of the S355MC-IN718 joint area: (**a**) iron (**b**) chromium (**c**) nickel (**d**) aluminium (**e**) titanium (**f**) niobium (**g**) molybdenum (**h**) manganese.

3.2. Finite Element Simulations

The results of the FE analysis are shown in Figure 11 using an S355MC-Inconel® 718 material pair. Maximum shear stress was selected as the parameter for study as it indicates the minimum required shear strength for the laminated joint. In Figure 11a, the simulation results show the maximum shear stress in the structure to be no greater than 37 MPa. This result defines the minimum shear strength required for the joining of two different materials used, and it will be compared against the experimental results. The maximum shear stress was selected for this comparison as it indicates the magnitude of the overall shear stress more clearly than shear stress in the *XZ*-plane under the current loading conditions. A clear difference in the maximum shear stress distribution of the two materials is seen in Figure 11a, as the non-magnetic Inconel® 718 is under almost two times greater shear stress than the magnetic S355 steel. This result can be explained by the greater density of Inconel® 718, and thus, a greater centrifugal load is experienced by this material.

In addition, it also possesses a lower modulus of elasticity, resulting in greater strain in Inconel® 718. Maximum principal stress is shown in Figure 11b. In this case, the magnetic S355 is under greatest stress having an approximate maximum principal stress of 77 MPa. This result indicates the minimum yield strength for the materials used. Additional stress results with different material pairs are listed in Table 7.

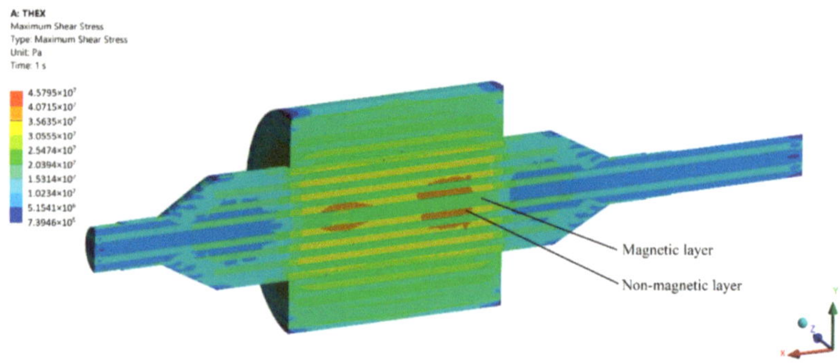

(**a**)

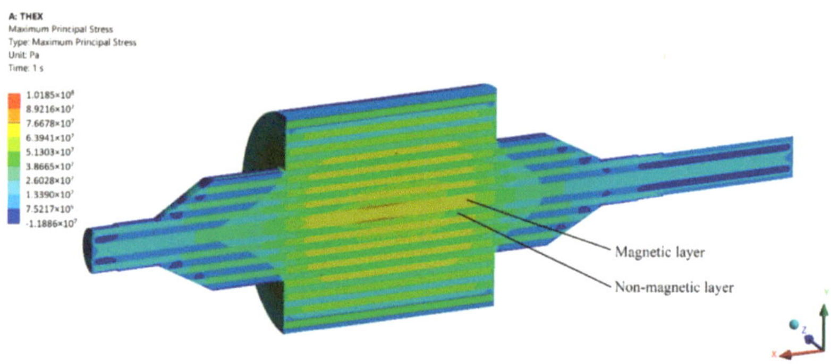

(**b**)

Figure 11. Stress distributions of laminated rotor structure under centrifugal and thermal loads. (**a**) Maximum shear stress distribution. (**b**) Maximum principal stress distribution.

Table 7. Stress results of rotor made of different material pairs.

Material Pair	Maximum Shear Stress (MPa)	Maximum Principal Stress (MPa)
(A) S355MC-Inconel® 600	59.4 (magnetic layer, center)	137.3 (non-magnetic layer)
(B) S1100MC-Inconel® 718	59.9 (non-magnetic layer)	116.5 (magnetic layer, center)
(C) S355MC-Inconel® 718	37.3 (non-magnetic layer)	76.61 (magnetic layer, center)

4. Discussion

None of the bonded test specimens A, B, and C achieved the same level of ultimate shear strength as the parent materials (Table 5). For S1100MC, the theoretical shear strength considering the von Mises yield criterion is about 790 MPa in its as-delivery state. With the same considerations, the shear strengths of Inconel® 600 and Inconel® 718 are about 400 MPa and 510 MPa, respectively. Therefore, the shear strength of the joints was lower than that of the parent materials, i.e., the measured shear strength of S355MCD (Table 5) and the theoretical shear strength of other parent materials in their delivery state. Moreover, it is noteworthy that the ultimate shear strength of the joint in S1100-IN718 (test specimen B) and S355-IN718 (test specimen C) is only 30 MPa higher in test specimen B although the strength of the parent material S1100 is much higher in the delivery state as compared to the S355 used in test specimen C.

In a study focused on describing the optimal vacuum brazing temperature of Inconel® 600 alloy when using AgCuTi braze alloy [25], it was stated that the shear strength reached the highest value of 223.32 MPa at a brazing temperature of 865 °C if Inconel® 600 is bonded by vacuum brazing. The study also drew attention to the profound effect of the process parameters, mainly brazing temperature, on the strength of the joint. It is noteworthy that the shear strength of 230.1 MPa of the vacuum brazed S355-IN600 joint shows that the shear strength of IN600 is not affected significantly by joining with S355 steel, although the braze alloy compositions and brazing process variables in the present study and in [25] are different. In [23], the ultimate tensile strength for a vacuum brazed S355J0-Inconel® 600 joint was found to be 283 MPa. Vacuum brazing Inconel® 718 was studied in [33] by examining γ-TiAl alloy/Inconel® 718 vacuum brazed joints. In the work, it is claimed that vacuum brazing IN718 is possible with a suitable braze alloy. The braze alloy used contained silver, copper, indium, and titanium, and the average shear strength achieved in the study was 228 $\pm$ 83 MPa at a brazing temperature of 730 °C.

The shear stress-normalized displacement curves (Figure 5) showed that there is some variation in the ductility of the joint since test specimen C shows different behavior compared to test specimens A and B. The behavior of test specimen C changes from linear elastic behavior to plastic region at lower shear stress. Overall, the ductility of test specimen C is higher than the ductility of test specimens A and B, as rupture occurred at a higher plastic strain. However, the ultimate shear strength of test specimen C is lower compared to other test specimens. These findings need thus to be interpreted with caution since this behavior could not be verified by actual measurement of the yield point with the test setup used in this study. The failure of the specimens observed in Figure 5 can be further confirmed by the study of the fracture surfaces (Appendix B). By observing the fractured specimens, it can be stated that test specimen A has failed by brittle fracture which is indicated by the very flat fracture surface. Test specimen B has failed by mixed fracture mode, which is noticeable by mildly slanted edges and partly flat fracture surface in the center of the specimen. Observing fracture surface of test specimen C, it can be stated that the failure mode is ductile failure due to the highly slanted edges of the fractured specimen.

The FE analysis described in Section 3 indicates a maximum permissible shear stress not to be greater than 37 MPa using the initial S355MC-Inconel® 718 material pair in the studied ALA-SynRM rotor structure at the nominal rotational speed and at operational temperature. Other material pair with S1100 and Inconel® 600 (material pairs A and B) yields greater shear stresses, although no greater than 60 MPa. It is noticeable in Table 7 that the location of maximum stresses changes between the magnetic center laminate layer and non-magnetic laminate layer with different material pairs, as illustrated for S355MC-Inconel® 718 pair in Figure 11. Magnetic steels S355 and S1100 have very similar material properties, as indicated in Table 4. The similar stress distributions in the laminate layers with material pairs B and C can be explained by the presence of Inconel® 718, as indicated in Table 7. In all material pairs, the material with a lower coefficient of thermal expansion is under the greatest tensile stress. Since the relative change in the density is smaller than the relative change in the coefficient of thermal expansion, it can be concluded that the main contribution to the simulated stresses is due to thermal strain. The shear strength values from the experimental tests support vacuum brazing and hot isostatic pressing methods as suitable bonding processes for ALA-SynRM rotors since both methods can achieve higher shear stress strength than the value predicted by FE analysis.

The microstructure and composition of the S355MC-IN600 vacuum brazed joint shows little or no diffusion of the main elements of the respective parent materials to their counterparts across the joint interface (Figure 8a,c). Additionally, a distinctive copper layer is present because of the copper braze alloy (Figure 8b). Manganese (Figure 8e) and silicon (Figure 8f) show good diffusion to the copper braze alloy. In addition, nickel shows some diffusion to the copper braze alloy (Figure 8c) whereas chromium (Figure 8d) and iron (Figure 8a) have slightly less diffusion to the copper braze alloy. A thin line of a nickel (Figure 8c) and chromium (Figure 8d) enriched area has also formed near the joint face of the S355MC. Additionally, on the joint face of IN600, precipitates of Cr and Cu have formed as indicated by the strong presence of chromium and copper in the same location in which dark spots are visible in the micrograph (arrow in Figure 8a). These Cr-Cu precipitates form alongside γ' (Ni_3Al, Ni_3Ti) precipitates, whose presence is confirmed by the precipitate-like morphologies seen in the Ni-rich region (arrow in Figure 8d).

In the S1100MC-IN718 joint, significant levels of aluminum, titanium, and niobium (Figure 9d–f), which are the main elements of minor phases, viz. γ' (Ni_3Al, Ni_3Ti) and γ'' (Ni_3Nb) precipitates [32], are seen along the joint face of IN718, indicating the formation of these precipitates. γ' and γ'' precipitates form in the joint area almost along a straight line as seen in Figure 9d–f. The dark precipitates in S1100MC near the joint could be MnS and SiC, based on the high intensity of Mn and Si visible in the respective composition plots (Figure 9h,i).

In the S355MC-IN718 joint, the main elements of each parent material show very little diffusion in the joint area (Figure 10a–c) and some aluminum, titanium, and niobium enriched areas are present in the form of γ' and γ'' precipitates. Γ' precipitates form along a straight line near the joint surface more prominently than the γ'' precipitates (Figure 10d–f).

The results in this study show that the long high temperature exposure during bonding processes affects the microstructure of the materials in a significant manner. The long exposure to temperatures above the recrystallization temperatures has caused changes in the microstructure and mechanical properties. The structural steels (S355MC, S1100MC) had reduced ultimate strength, whereas the ultimate strength of the Inconel® grades was higher than the values specified by the manufacturers in the delivery state (Table 6).

All materials have exhibited grain growth due to the high temperature exposure. For example, for Inconel® 600, grain growth begins at a temperature of 980 °C by coalescing carbides in the microstructure. Exposure for 1–2 h at temperatures of 1090–1150 °C will completely dissolve carbides and result in an increased grain size [28].

As the HIP process is carried out at 1150 °C for 5 h (Table 2), high strength phases such as martensite and bainite revert to austenite, which then transforms to ferrite and pearlite during the slow cooling of the joint. Moreover, at such high temperatures, significant grain growth occurs. The sharp drop in hardness of S1100MC after the joining process (Figure 6b) is attributed to the large austenite grain sizes and change in phase composition, i.e., from high strength and hard phases such as martensite and bainite to a relatively soft coarse-grained ferrite–pearlite mixture. Despite the loss of the lath martensitic structure due to the long-time high temperature exposure during HIP, the shear strength of the S1100MC-IN718 joint is slightly higher than the shear strength of the S355MC-IN718 joint and almost the same as the strength of the S355MC-IN600 joint.

Although the HIP process temperatures for S1100-IN718 and S355-IN718 joints are the same (1150 °C), the dwell time of the S1100-IN718 joint is longer compared to the other joint (Table 2). This longer process time used for the S1100-IN718 joint enables the diffusion of alloying elements, such as Al, Nb, Ti, much closer towards the joint region. This gives rise to the formation and concentration of γ' and γ'' precipitates very near to the joint region rather than in regions that are slightly farther away from the joint (Figure 9). Consequently, this leads to depletion of alloying elements (Ti, Al, Nb) and a lack of evenly distributed precipitates in the regions that are slightly farther away from the joint, thereby giving rise to the observed sudden variations in the hardness of IN718 near the joint (Figure 6b). In the S355-IN718 joint, due to a relatively lower dwell time, the alloying elements, such as Al, Ti, Nb are more evenly distributed in IN718, consequently leading to even distribution of the precipitates (Figure 10d–f). Therefore, no sudden drop in hardness near the joint is observed (Figure 6c).

5. Conclusions

In this work, dissimilar metal joints which can be used to bond alternating magnetic and non-magnetic metal sheets to manufacture a high-speed ALA-SynRM rotor were studied. The material pairs and bonding types selected for study were vacuum-brazed S355MCD-Inconel® 600 and hot isostatically pressed S1100MC-Inconel® 718, and S355MC-Inconel® 718. Shear stress tests were performed to determine the shear strength of the joints. Additionally, finite element analysis (FEA) of the ALA-SynRM rotor was performed to predict the required strength of the joints at the nominal rotation speed of the proposed rotor geometry. FEA showed that the material parameters (density, elastic modulus, and thermal expansion ratio) affect the stresses that the rotor experiences during operation. Material pair S355-IN718 generated the lowest stresses in this study. FEA showed that the joints should have a shear strength of at least 60 MPa, 60 MPa, and 37 MPa for S355-IN600, S1100-IN718, and S355-IN718, respectively. FEA also showed that the parent materials should have a minimum yield stress of at least 137 MPa, 117 MPa, and 77 MPa in S355-IN600, S1100-IN718, and S355-IN718, respectively.

In the experimental tests, the hot isostatic pressed S1100MC-IN718 joint achieved the highest ultimate shear strength (233.3 MPa) followed by the vacuum-brazed S355MCD-IN600 joint (230.1 MPa) and the HIP S355-IN718 (203.5 MPa) joint. It is noteworthy that the strength of ultra-high strength structural steels, like S1100MC, cannot be utilized when using bonding processes that require long exposure to high temperature as the heat input affects the microstructure of the parent material. However, the loss in strength and hardness of S1100MC after bonding does not seem to affect the overall ultimate shear strength of the S1100MC-IN718 joint compared to that of the S355MCD-IN600 joint. In conclusion, all test specimens of the study meet the requirements set by the operation conditions and geometry of the studied ALA-SynRM rotor. Therefore, it can be concluded that vacuum brazing and hot isostatic pressing can be considered viable bonding methods to fabricate high-speed ALA-SynRM rotors.

Author Contributions: Conceptualization, E.S. (Eero Scherman); Data curation, E.S. (Eero Scherman) and E.S. (Eerik Sikanen); Funding acquisition, J.S.; Investigation, E.S. (Eero Scherman); Methodology, E.S. (Eero Scherman), H.K.Y. and M.A.; Project administration, J.S.; Resources, E.S. (Eero Scherman); Software, E.S. (Eerik Sikanen); Supervision, J.S.; Validation, H.K.Y. and J.S.; Visualization, E.S. (Eero Scherman) and E.S. (Eerik Sikanen); Writing—original draft, E.S. (Eero Scherman), E.S. (Eerik Sikanen), H.K.Y. and M.A.; Writing—review & editing, H.K.Y. and J.S. All authors have read and agreed to the published version of the manuscript.

Funding: This research was funded by Business Finland, grant numbers 5648/31/2018, 5649/31/2018, and partly by INERCOM (Integrated Energy Conversion Machinery)—a platform of LUT University.

Institutional Review Board Statement: Not applicable.

Informed Consent Statement: Not applicable.

Data Availability Statement: Data are contained within the article "Microstructure and mechanical properties of steel and Ni-based superalloy joints for rotors of high-speed electric motors.".

Conflicts of Interest: The authors declare no conflict of interest.

Appendix A

Table A1. Chemical composition (weight %) of the parent materials based on material certificates.

Material	Thickness (mm)	C	S	Cr	Ni	V	Mo	Mn	N	Si	Ti	Nb	Cu	Fe	P	Al	B	Co	Ta	Bi	Pb	Ag
S355MC	3.00	0.0686	0.0041	0.02332	0.0292	0.0047	0.0018	0.5371	0.0052	0.0147	0.0010	0.0218	0.0094	Bal	0.0122	0.0355	0.0001	-	-	-	-	-
S355MCD	6.00	0.060	0.005	-	-	0.006	-	0.69	-	0.01	0.001	0.026	-	Bal	0.010	0.036	-	-	-	-	-	-
S1100MC	4.00	0.141	0.001	0.69	0.30	0.06	0.05	1.40	0.004	0.19	0.01	0.001	0.01	Bal	0.010	0.033	0.0011	-	-	-	-	-
S1100MC	6.00	0.140	0.001	0.68	0.29	0.06	0.06	1.41	0.003	0.32	0.01	0.001	-	Bal	0.011	0.034	0.0014	-	-	-	-	-
Inconel®600	3.00	0.07	0.002	16.13	73.58	-	-	0.21	-	0.33	0.20	0.01	0.03	9.11	0.006	0.18	-	0.02	0.01	-	-	-
Inconel®718	3.17	0.05	0.0002	18.16	52.81	-	-	0.05	-	0.06	1.02	-	0.04	18.99	0.008	0.58	0.002	0.12	0.01	-	-	-
Inconel®718	3.71	0.03	0.0003	18.28	53.47	-	-	0.09	-	0.08	0.96	5.06	0.06	18.2	0.007	0.51	0.001	0.17	0.01	-	-	-
CW021A	0.100	-	-	-	-	-	-	-	-	-	-	-	99.987	-	0.0050	-	-	-	-	0.0001	0.0001	0.0150

Table A2. Material manufacturers and origins.

Material	Thickness (mm)	Company	Country
S355MC	3.00	ArcelorMittal Bremen Gmbh	Germany
S355MCD	6.00	SSAB Europe Oy	Finland
S1100MC	4.00	SSAB Emea AB	Sweden
S1100MC	6.00	SSAB Emea AB	Sweden
Inconel®600	3.00	VDM Metals Gmbh	Germany
Inconel®718	3.17	ATI Specialty Rolled Products	United States of America
Inconel®718	3.71	Huntington Alloys Corporation	United States of America
CW021A	0.100	Ferd. Haecker Gmbh & CO. KG	Germany

Appendix B

Figure A1. Fracture surfaces of test specimens.

References

1. Hughes, A.; Drury, B. Chapter 1—Electric Motors—The Basics. In *Electric Motors and Drives*, 5th ed.; Hughes, A., Drury, B., Eds.; Newnes: Oxford, UK, 2019; pp. 1–39. [CrossRef]
2. Gerada, D.; Mebarki, A.; Brown, N.L.; Gerada, C.; Cavagnino, A.; Boglietti, A. High-Speed Electrical Machines: Technologies, Trends, and Developments. *IEEE Trans. Ind. Electron.* **2014**, *61*, 2946–2959. [CrossRef]
3. Smirnov, A.; Uzhegov, N.; Sillanpää, T.; Pyrhönen, J.; Pyrhönen, O. High-Speed Electrical Machine with Active Magnetic Bearing System Optimization. *IEEE Trans. Ind. Electron.* **2017**, *64*, 9876–9885. [CrossRef]
4. Tenconi, A.; Vaschetto, S.; Vigliani, A. Electrical Machines for High-Speed Applications: Design Considerations and Tradeoffs. *IEEE Trans. Ind. Electron.* **2014**, *61*, 3022–3029. [CrossRef]
5. Moghaddam, R.R. High Speed Operation of Electrical Machines, a Review on Technology, Benefits and Challenges. In Proceedings of the 2014 IEEE Energy Conversion Congress and Exposition (ECCE), Pittsburgh, PA, USA, 14–18 September 2014; pp. 5539–5546.
6. Arkkio, A.; Jokinen, T.; Lantto, E. Induction and Permanent-Magnet Synchronous Machines for High-Speed Applications. In Proceedings of the 2005 International Conference on Electrical Machines and Systems, Sofia, Bulgaria, 27–29 September 2005; Volume 2, pp. 871–876.
7. Abramenko, V.; Petrov, I.; Nerg, J.; Pyrhönen, J. Synchronous Reluctance Motors With an Axially Laminated Anisotropic Rotor as an Alternative in High-Speed Applications. *IEEE Access* **2020**, *8*, 29149–29158. [CrossRef]
8. Tsao, P.; Senesky, M.; Sanders, S. A Synchronous Homopolar Machine for High-Speed Applications. In Proceedings of the Conference Record of the 2002 IEEE Industry Applications Conference. 37th IAS Annual Meeting (Cat. No.02CH37344), Pittsburgh, PA, USA, 13–18 October 2002; Volume 1, pp. 406–416.
9. Heidari, H.; Rassõlkin, A.; Kallaste, A.; Vaimann, T.; Andriushchenko, E.; Belahcen, A.; Lukichev, D.v. A Review of Synchronous Reluctance Motor-Drive Advancements. *Sustainability* **2021**, *13*, 729. [CrossRef]
10. Kurvinen, E.; Di, C.; Petrov, I.; Jastrzebski, R.P.; Kepsu, D.; Pyrhönen, J. Comparison of the Performance of Different Asynchronous Solid-Rotor Constructions in a Megawatt-Range High-Speed Induction Motor. In Proceedings of the 2019 IEEE International Electric Machines & Drives Conference (IEMDC), San Deigo, CA, USA, 12–15 May 2019; pp. 820–825.
11. ASM Handbook. *Welding, Brazing and Soldering*; Olson, D.L., Siewert, T.A., Liu, S., Edwards, G.R., Eds.; ASM International: New York, NY, USA, 1993.
12. Li, Y.; Chen, C.; Yi, R.; Ouyang, Y. Review: Special Brazing and Soldering. *J. Manuf. Process.* **2020**, *60*, 608–635. [CrossRef]
13. Costanza, G.; Tata, M.E.; Cioccari, D. Explosion Welding: Process Evolution and Parameters Optimization. *Mater. Sci. Forum* **2019**, *941*, 1558–1564. [CrossRef]
14. Bocanegra-Bernal, M.H. Hot Isostatic Pressing (HIP) Technology and Its Applications to Metals and Ceramics. *J. Mater. Sci.* **2004**, *39*, 6399–6420. [CrossRef]
15. Guo, J. Solid State Welding Processes in Manufacturing. In *Handbook of Manufacturing Engineering and Technology*; Nee, A.Y.C., Ed.; Springer: London, UK, 2015; pp. 569–592. ISBN 978-1-4471-4670-4.
16. Beser, E.; Camur, S.; Arifoglu, B.; Beser, E. Design and Analysis of an Axially Laminated Reluctance Motor for Variable-Speed Applications. *Adv. Electr. Comput. Eng.* **2013**, *13*, 75–80. [CrossRef]
17. Kim, Y.; Lee, J. Optimum Design Criteria of an ALA-SynRM for the Maximum Torque Density and Power Factor Improvement. *Int. J. Appl. Electromagn. Mech.* **2017**, *53*, 1–10. [CrossRef]
18. Saxena, R.; Banerji, G.K.; Srivastava, A.; Rawat, H.S. Performance Analysis of Axially Laminated Anisotropic Synchronous Reluctance Motor. In Proceedings of the 7th WSEAS International Conference on Electric Power Systems, High Voltages, Electric Machine, Venice, Italy, 21–23 November 2007; pp. 97–102.
19. Veg, L.; Laksar, J. Overview of Technology, Problems and Comparison of High Speed Synchronous Reluctance Machines. In Proceedings of the 2018 ELEKTRO, Mikulov, Česko, 21–23 May 2018; pp. 1–4.
20. Hofmann, H.; Sanders, S.R. High-Speed Synchronous Reluctance Machine with Minimized Rotor Losses. *IEEE Trans. Ind. Appl.* **2000**, *36*, 531–539. [CrossRef]

21. Hausmann, L.; Waldhof, M.; Fischer, J.; Wößner, W.; Flammer, M.O.; Heim, M.; Fleischer, J.; Parspour, N. Review and Enhancements of Rotor Designs for High Speed Synchronous Reluctance Machines. In Proceedings of the 2020 10th International Electric Drives Production Conference (EDPC), Ludwigsburg, Germany, 8–9 December 2020; pp. 1–8. [CrossRef]
22. Kim, Y.H.; Lee, J.H. Optimum Design of ALA-SynRM for Direct Drive Electric Valve Actuator. *IEEE Trans. Magn.* **2017**, *53*, 8200804. [CrossRef]
23. Abramenko, V.; Nerg, J.; Petrov, I.; Pyrhönen, J. Influence of Magnetic and Nonmagnetic Layers in an Axially Laminated Anisotropic Rotor of a High-Speed Synchronous Reluctance Motor Including Manufacturing Aspects. *IEEE Access* **2020**, *8*, 117377–117389. [CrossRef]
24. Lee, H.-S. 10—Diffusion Bonding of Metal Alloys in Aerospace and Other Applications. In *Welding and Joining of Aerospace Materials*, 2nd ed.; Chaturvedi, M., Ed.; Woodhead Publishing: Sawston, UK, 2021; pp. 305–327. [CrossRef]
25. Zaharinie, T.; Yusof, F.; Hamdi, M.; Ariga, T.; Moshwan, R. Effect of Brazing Temperature on the Shear Strength of Inconel 600 Joint. *Int. J. Adv. Manuf. Technol.* **2014**, *73*, 1133–1140. [CrossRef]
26. Special Metals Corporation INCONEL®Alloy 718. Publication Number SMC-045 2007. Available online: https://www.specialmetals.com/documents/technical-bulletins/inconel/inconel-alloy-718.pdf (accessed on 16 December 2021).
27. Ovako Steel Grade S355JR. Material Data Sheet 2020. Available online: https://steelnavigator.ovako.com/steel-grades/s355/pdf (accessed on 16 December 2021).
28. Special Metals Corporation INCONEL®Alloy 600. Publication Number SMC-027 2008. Available online: https://www.specialmetals.com/documents/technical-bulletins/inconel/inconel-alloy-600.pdf (accessed on 25 December 2021).
29. SSAB Ab. *Design Handbook Structural Design and Manufacturing in High-Strength Steel*, 1st ed.; SSAB Ab: Borlänge, Sweden, 2017.
30. Hornbuckle, B.C.; Murdoch, H.A.; Roberts, A.J.; Kecskes, L.J.; Tschopp, M.A.; Doherty, K.J.; Yu, J.H.; Darling, K.A. Property Mapping of Friction Stir Welded Al-2139 T8 Plate Using Site Specific Shear Punch Testing. *Mater. Sci. Eng. A* **2017**, *682*, 192–201. [CrossRef]
31. Afkhami, S.; Javaheri, V.; Amraei, M.; Skriko, T.; Piili, H.; Zhao, X.-L.; Björk, T. Thermomechanical Simulation of the Heat-Affected Zones in Welded Ultra-High Strength Steels: Microstructure and Mechanical Properties. *Mater. Des.* **2022**, *213*, 110336. [CrossRef]
32. Sonar, T.; Balasubramanian, V.; Malarvizhi, S.; Venkateswaran, T.; Sivakumar, D. An Overview on Welding of Inconel 718 Alloy—Effect of Welding Processes on Microstructural Evolution and Mechanical Properties of Joints. *Mater. Charact.* **2021**, *174*, 110997. [CrossRef]
33. Sequeiros, E.W.; Guedes, A.; Pinto, A.M.P.; Vieira, M.F.; Viana, F. Microstructure and Strength of γ-TiAl Alloy/Inconel 718 Brazed Joints. *Mater. Sci. Forum* **2013**, *730–732*, 835–840. [CrossRef]

materials

MDPI

Article

Microstructural and Performance Analysis of TP304H/T22 Dissimilar Steel Welded Joints

Jian Sun [1,2,*], Tong Wang [1], Fuguang Liu [3], Zhoubo Zhang [3], Yunhui Chen [1], He Lin [1,2], Hui Liu [1,2], Xiaohui Zhao [1,2] and Xiaole Cheng [1,2]

1 School of Mechanical and Electrical Engineering, Xi'an Polytechnic University, Xi'an 710048, China
2 Xi'an Key Laboratory of Modern Intelligent Textile Equipment, Xi'an 710048, China
3 Xi'an Thermal Power Research Institute Co., Ltd., Xi'an 710054, China
* Correspondence: sunjian@xpu.edu.cn

Abstract: In the power plant boiler industry, dissimilar steel welding is widely used in the connection of thermal power generation units. As an important component of the unit, research on the organizational properties of dissimilar steel welded joints has significant guidance for the life design of the joint. For the long-term service state of TP304H/T22 dissimilar steel welded joints, the microstructure's morphological evolution, the microhardness, and the tensile properties of tube samples were analyzed using tests and numerical simulations. The results show that the microstructure of each part of the welded joint was free of damaged features, such as a creep cavity and intergranular cracks. The microhardness of the weld was higher than that of the base metal. In the tensile test, the welded joints broke at the weld metal at room temperature and at the side of the TP304H base metal at a temperature of 550 °C. The tensile fracture morphology demonstrated a change from a ductile fracture to a hybrid fracture when the temperature rose. The fusion zone and base metal on the TP304H side were the stress concentration areas of the welded joint, which easily sprouted cracks. This study holds significant reference value in assessing the safety and reliability of dissimilar steel welded joints in superheater units.

Keywords: dissimilar welded joints; microstructure; hardness; tensile properties

check for updates

Citation: Sun, J.; Wang, T.; Liu, F.; Zhang, Z.; Chen, Y.; Lin, H.; Liu, H.; Zhao, X.; Cheng, X. Microstructural and Performance Analysis of TP304H/T22 Dissimilar Steel Welded Joints. *Materials* **2023**, *16*, 4474. https://doi.org/10.3390/ma16124474

Academic Editors: Swarup Bag and Christ Prakash Paul

Received: 25 April 2023
Revised: 8 June 2023
Accepted: 9 June 2023
Published: 20 June 2023

1. Introduction

Dissimilar material welding is widely used in various fields because, in some cases, we need to use different types of materials and must join them together to form a complete structure. When conducting dissimilar material welding, the most common welding material used is dissimilar steel, and because of the different coefficients of thermal expansion, chemical compositions, physical properties, and other significant differences of the steel, the welded dissimilar steel is much more complex than when the same steel is welded. Problems associated with dissimilar steel welding, such as carbon migration, poor metallurgical bonding in the fusion zone, relatively complex stress fields at joints, microstructural deterioration, and the presence of cracks near the weld, severely affect the quality and safety of the welded joint [1–4]. TP304H and T22 are common dissimilar steel welding combinations in production. In thermal power generating units, many welded structures are manufactured using welded austenitic stainless-steel and low-alloy steel dissimilar metals [5–7]. Such welded joints operate under different temperatures and pressures for long periods, which necessitates the requirements of high temperature stability, high pressure bearing capacity, and chemical stability. The main problem associated with this kind of dissimilar steel welded joint is that the service life of the joint is shorter than the design life, and failure is mainly due to the low plastic cracking along the weld interface [2]. This kind of failure is usually sudden, and it easily causes damage to the welded structure, affecting normal production and resulting in huge economic losses. It is of great significance to study the microstructure and properties of dissimilar steel welded joints to ensure the safe and stable operation of units.

In the dissimilar steel welding process, because of changes to the metallurgical organization of the welded area or the generation of new metallurgical organizational arrangements, the performance of welded joints usually degrades. In order to evaluate the quality of welded joints, the microstructures of each zone of welded joints were obtained by optical and scanning electron microscopy, and the reasons for the different microstructures of dissimilar welded joints were studied [3]. Zhang [8] characterized semicircular and linear layered textures in fusion zone and tensile fracture features based on the microstructure. Ou [9] paid more attention to the evolutionary process of the microstructural morphology of the welded joint and obtained the structure–property relationship of the joint. Zhai [10] further investigated the effect of the microstructural evolution on the creep properties. Mittal [11] selected different combinations of welding processes and welding materials and evaluated the performance of the joint based on the microhardness and fracture results. The failure mode of most of the welded structures was brittle fracture under low stress levels, with little apparent plastic deformation before fracture. Battahar [12] found that the presence of macroscopic defects or the location of cracks in mechanical structures was the main source of fatigue crack initiation. Comprehensive studies by Wang [13] and Chen [14] showed that the mechanism of crack initiation and propagation exists mainly in the fusion line region, as shown in different microstructures in dissimilar welded joints. Sun [15] mainly investigated the effects of chemical composition and grain size on near-threshold fatigue crack growth behavior. In addition, some research focused on adjusting the strength–toughness matching of welded joints. Aghajani Derazkola [16] analyzed the formation mechanism of intermetallic compounds during dissimilar welding. Chen [17] increased the tensile strength of dissimilar welded joints by reducing the content of intermetallic compounds. Fan [2] analyzed failed dissimilar steel tube samples and found that the stress concentration effect was the main reason for the cracking of the dissimilar steel weld before the design life. For the study of the stress distribution pattern in welded joints, Zhang [18] obtained the residual stress distribution in adjacent welded joints through numerical simulation methods. Adomako [19] discussed the deformation mechanism of dissimilar welded joints and predicted the tensile behavior of the welded joint. Therefore, the early failure of welded joints can be avoided by analyzing the evolution of the microstructure and diffusion mechanism of the elements, paying attention to the location of the area of the stress concentration in the welded joints. The variation law of the temperature and stress field during the welding process can be obtained, and the residual stress and deformation behavior of welded joints can be predicted through the numerical simulation of welded joints. Sun [20] discussed the crack propagation path based on the extended finite element method. The results of the study confirmed the effectiveness of numerical simulation for predicting the expansion of crack emergence in welded joints. The feasibility of using numerical simulation to optimize the welding parameters and the reliability of the mechanical property analysis of welded joints based on numerical simulation results has been verified in the literature [21–25]. Therefore, based on previous studies, experimental and numerical simulation methods were used in this study for predicting mechanical properties across dissimilar material welded joints.

In this study, a welded joint of TP304H/T22 dissimilar steel, which has served for a long time at the outlet of a secondary superheater of a 350 MW subcritical boiler, was taken as the research object. The mechanical properties of the welded joint were analyzed. The microstructure, microhardness, and tensile properties of each region of the welded joint were studied using experiments. Additionally, the finite element method was utilized to analyze the high-temperature tensile process of the welded joint. According to the stress distribution law of the welded joint and the evolution of the microstructure of the welded joint, the weak areas of the welded joint were obtained. This study provides a reference for the reliability of dissimilar steel welded joints under long-term high-temperature and high-pressure operation and the avoidance of early failure phenomena such as cracking.

2. Experimental Materials and Methods

The TP304H/T22 dissimilar steel welded joint of a 350 MW subcritical boiler was taken as the research object. The dissimilar steel welded joint was located on the roof of the outlet section of the boiler superheater, and it had been continuously operated under high-temperature and high-pressure conditions over a long period of time. The transverse cross-sectional morphology of the welded joint is shown in Figure 1. The TP304H and T22 tube samples had diameters of $\Phi 50.8 \times 7.4$ mm and $\Phi 50.8 \times 12.2$ mm, respectively. The chemical compositions are shown in Tables 1 and 2. Because of the different wall thickness of the base metal on both sides, the dissimilar steel weld added a backing plate (T22). The contact position between the backing plate and the base metal on both sides was not welded together. A nickel-based alloy welding wire was selected as the welding material for the dissimilar steel joints. The welding wire contained high amounts of Ni and Cr elements and a certain amount of alloy elements, such as Nb, Mo, and Ti. Its linear expansion coefficient was between pearlitic steel and austenitic steel, which can reduce the generation of thermal stress, prevent the diffusion and migration of carbon near the fusion zone, and ensure the weldability and performance of the joint [5]. The metallographic structure of the welded joint of TP304H/T22 dissimilar steel was observed and analyzed with a ZEISS IMAGER A1m metallographic microscope, and the metallographic structure of the base metal is shown in Figure 2. It can be seen that the TP304H is a single austenite structure, and the T22 consists of ferrite, pearlite, and granular bainite with uniform distribution. The two dissimilar materials, TP304H and T22, are metallurgically compatible and easily form new phase structures or metal compounds during the welding process. They have a certain weldability and form good, welded joints.

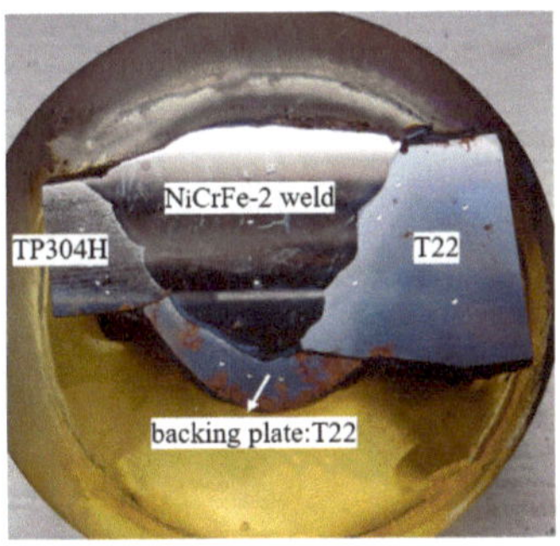

Figure 1. Section morphology of the welded joint.

Table 1. Chemical compositions of T22 (mass fraction, %).

Material	Cr	Mo	Mn	Si	C	S	P	Fe	Ni	Nb
T22	2.08	0.92	0.44	0.22	0.088	0.016	0.0098	Balance	-	-

Table 2. Chemical compositions of TP304H (mass fraction, %).

Material	Cr	Ni	Mn	Si	C	P	S	Fe	Mo	Nb
TP304H	18.82	9.82	1.52	0.55	0.045	0.026	0.012	Balance	-	-

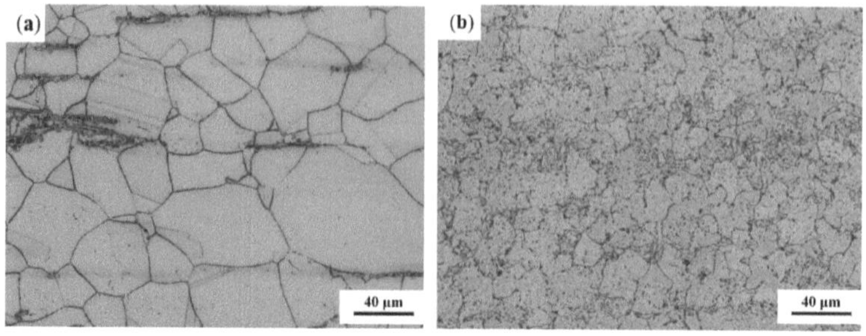

Figure 2. Microscopic morphology of the base metal: (**a**) TP304H; (**b**) T22.

The microstructure of the welded joint was observed using a Thermo Scientific Apreo field-emission scanning electron microscope, and the second phase composition was analyzed with X-ray energy-dispersive spectrometry (EDS). Specimens were prepared according to standard metallographic procedures for steel, including embedding, grinding, polishing, and solution etching, prior to observing the microstructure. A digital display Vickers hardness tester was used to measure the Vickers microhardness on the specimen surface at a fixed spacing of 1 mm. The rectangular tensile specimens with full wall thickness were prepared and tested using an electronic universal tensile testing machine at room temperature and 550 °C.

3. Microstructure of Welded Joints

In the dissimilar steel joint tube sample, a small amount of the precipitated phases and inclusions in the base metal zone on the TP304H side were observed. Unevenly sized equiaxial twins connected, and the austenite grains of the base metal at the melting boundary directly contacted the welded molten pool cell to generate epitaxial nucleation, as shown in Figure 3a. In addition, the homogenization caused by solid diffusion was small due to the matrix of the fusion zone on the TP304H side still being austenite and the structure being tightly arranged [9]. Close to the fusing line side, the austenite grains grew epitaxially from the base metal to the weld recrystallization zone.

The microstructure of the heat-affected zone on the T22 side is shown in Figure 3b. There were more evenly dispersed massive ferrite and a small amount of lathy ferrite. In this area, the bainite morphology essentially disappeared and dispersed carbides precipitated at the grain boundary. This was due to the existence of a large amount of chromium in the T22 matrix, which caused the carbide to form a grain structure at high temperature. Compared to the metallographic organization of the T22 base metal, the grain refinement in the heat-affected zone had a better ability to resist crystallization cracks. In addition, as the same material was used between the backing plate and the base metal, the welding and fusion performance improved, and no damage characteristics, such as creep cavity and intergranular cracks, were found, avoiding the fracture of joints caused by crystal cracks as the source of fatigue cracks.

Figure 3c shows the microstructure of the center area of the weld. The weld metal composition phase was mainly austenite with coarse grains, large spacing between dendrite arms and cell crystals, and small chrome-rich carbides precipitated at the grain boundary, among which some oxides were included to produce shrinkage holes [6].

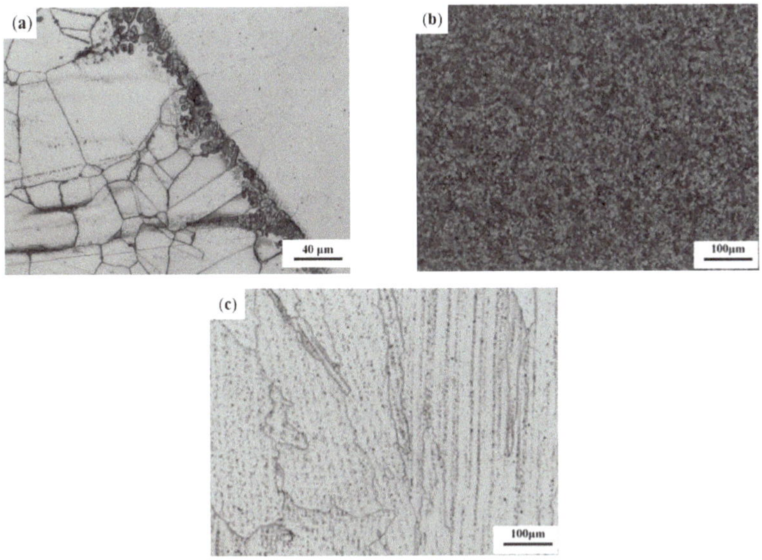

Figure 3. Microstructure of the welded joint: (**a**) TP304H side adjacent to the fusion line; (**b**) T22 side heat-affected zone; (**c**) weld.

As shown in Figure 4a, the weld had white precipitates distributed in and at the grain boundary, which were analyzed using EDS. The EDS spectrum results are shown in Figure 4b. The mass fractions of the different elements are shown in Table 3. The elements mainly detected include Fe, Cr, and Nb. Nb easily combined with C and N to form the Nb (C, N) phase, which was diffusely distributed within the inner crystal and at grain boundaries [26]. This was beneficial for improving the strength and impact toughness of the welded joint. At the measurement point 02, the elements O, Al, and Ti were mainly detected. Refer to the research results of other similar literature, where the higher oxygen content in the weld metal could cause oxidative corrosion effects, which could lead to cracks in the joint during future service processes [27]. Combined with Figure 3c, the microstructure of the weld showed that the weld grain boundary precipitated tiny carbides, and there was a tendency to interconnect into a chain, and the second phase particles with small and many fine sizes inhibited the growth of grains by pinning grain boundaries [15], which reduced the size of previously generated austenite grains and easily induced cracks at the weld.

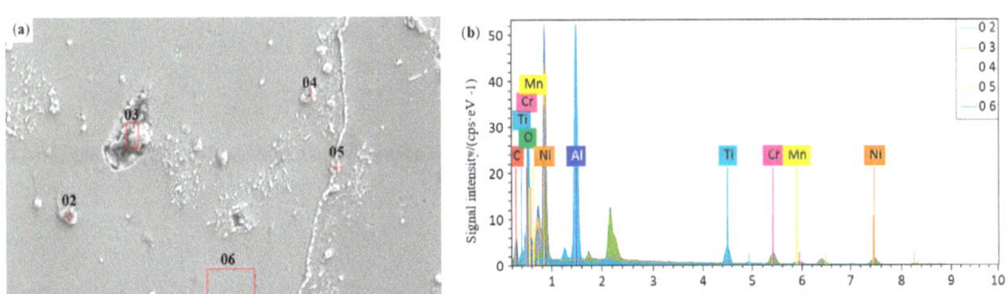

Figure 4. Weld precipitation phase: (**a**) location distribution of EDS measurement points; (**b**) EDS test results.

Table 3. Normalized mass percentages of elements at different measurement point locations (%).

Locations	C	O	Al	Si	Ti	Cr	Mn	Fe	Ni	Nb	Hg
02	9.55	34.97	26.94	-	18.38	2.22	1.58	-	6.35	-	-
03	7.33	-	-	2.34	2.69	15.41	3.31	10.43	26.25	32.23	-
04	5.52	-	-	0.98	-	16.51	4.45	18.51	50.66	3.37	-
05	9.03	-	-	0.51	-	14.23	3.72	17.26	53.86	1.40	-
06	2.75	-	-	0.49	-	16.30	2.29	18.32	58.53	-	1.32

4. Mechanical Properties of Welded Joints

4.1. Hardness

The micro-Vickers hardness test was performed on the tube samples with a test load of 10 kgf and a holding time of 12 s. The tube sample was utilized for microhardness testing. The hardness value measurement numbering method started from the fusion boundary R onwards to the weld or the direction of the base metal. M was used to indicate the distribution of the measurement points on the base metal, H to indicate the measurement points on the weld, and R to indicate the measurement points in the heat-affected zone. The number represents the serial number of the measurement points, with each measurement point having an interval of 1 mm. The Vickers hardness value test results are shown in Figure 5. The average hardness of the TP304H base metal was 172 HV, and the average hardness of the T22 base metal was 132 HV. According to DL/T 438-2016 "*Technical Supervision Regulations for Metals in Thermal Power Plants*", the hardness of TP304H is required to be 147–202 HV, and the hardness of T22 is required to be 131–189 HV. The weld zone was made of NiCrFe-2, which has good strength and plasticity at high temperature. The test results show that the hardness of the base metal on both the T22 and TP304H sides met the standards, with the average hardness of T22 being close to the lower limit of the standard and the average hardness of TP304H being moderate.

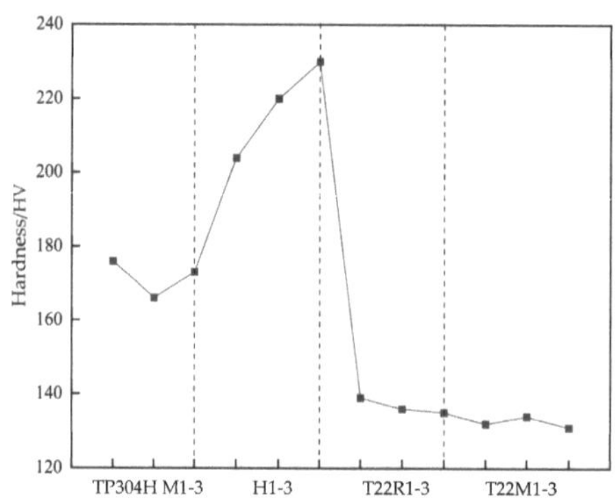

Figure 5. Vickers hardness distribution of the TP304H/T22 dissimilar welded joint.

As shown in Figure 5, the hardness values of the specimen in the TP304H base metal zone were similar to those in the fusion line, indicating that no phenomena such as interdiffusion and new phase generation occurred between the two parts. The chemical composition, as shown in Table 2, and the microstructure, as shown in Figure 3a, further prove this conclusion. The average value of the hardness of the specimen in the TP304H base metal was lower than the average value of the hardness in the weld. The hardness of the specimen increased abruptly in the weld zone due to the hindrance of the dislocation slip of the grain boundary in the heat-affected zone on the TP304H side (Figure 3a). The difference in the hardness values between the weld and the base metal might lead to joint cracking. The heat-affected zone on the T22 side had higher hardness values than the T22 base metal, with an average hardness value of 137 HV. The metallographic organization of the heat-affected zone on the T22 side (Figure 3b) showed that the grain size in the heat-affected zone was smaller compared to the base metal, and the hardness at this location increased because of the fine grain strengthening effect. The maximum hardness of the welded heat-affected zone also reflects the strength of the welded heat-affected zone. The higher the strength, the poorer the toughness and plasticity, and the easier it is to produce or extend cracks [11]. The lower hardness value of the base metal on the T22 side indicates that there is some degree of damage to the pipe during use, and the decrease in the hardness value is mainly caused by carbide spheroidization [28].

4.2. Tensile Properties at Different Temperatures

Four samples at the same position on the pipe sample were cut with the weld seam as the center to make full-thickness rectangular tensile samples, numbered 1, 2, 3, and 4. The tensile properties of the welded joints at room temperature (RT) and 550 °C are shown in Table 4, and the stress–strain curves are shown in Figure 6. Specimens 1 and 2 were high-temperature tensile specimens, and specimens 3 and 4 were room-temperature tensile specimens. The high-temperature tensile specimens were fractured at different locations on the TP304H side. Specimen 1 was fractured on the TP304H side of the base metal, indicating that the strength of the joint was higher than that of the carbon steel base metal. The tensile fracture morphology was rosy red, the fracture plane was at an angle of approximately 45° to the tensile axis, and the fracture was wedge shaped. Specimen 2 was broken along the fusion line on the TP304H side, and the fracture morphology was also rosy red. Bai conducted high-temperature tensile tests on unserved joints, and the results showed that all joints fractured on the low-alloy side [7]. In this study, high-temperature tensile tests were performed on welded joints after long-term service, which fractured on the austenite side. All room-temperature tensile specimens fractured at the weld on the TP304H side. The fracture morphology was gray–white, and the plastic deformation at the fracture site was large.

Table 4. Tensile properties of the TP304H/T22 welded joint specimens.

Specimen	Tensile Strength (MPa)	Standard Tensile Strength (GB/T 5310-2017) (MPa)	Fracture Positions
1 (HT)	389	≥381 (TP304H at high temperature)	TP304H side base metal
2 (HT)	334		TP304H side fusion line
3 (RT)	476	≥515 (TP304H at room temperature)	Welding seam
4 (RT)	330	≥415 (T22 at room temperature)	Welding seam

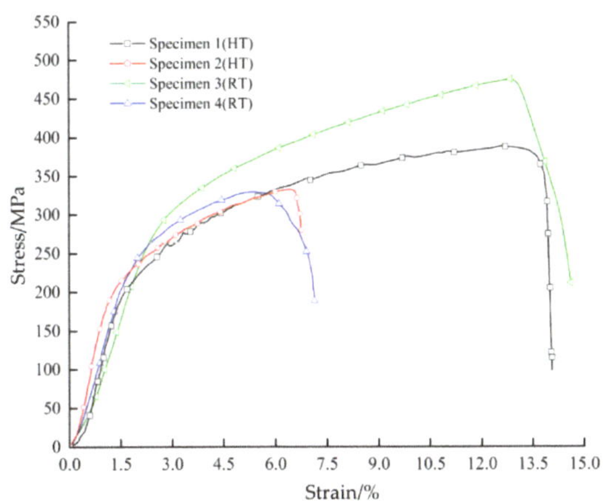

Figure 6. Tensile stress–strain curves of the TP304H/T22 dissimilar welded joint.

Compared with the fracture positions of the tensile test, the plastic deformation of the welded joints at 550 °C was significantly lower than the deformation at room temperature. This was because the dislocation slip of the grains was prone to occur at high temperatures. In addition to the cross-slip of the spiral dislocation, new dislocation was activated by stress, leading to dynamic recovery and softening [8]. The austenitic–low-alloy steel dissimilar welded joints were used as pressure-bearing parts of the thermal power generating units. At some interfaces and microcracks inside the crystal, thermal stress and tissue stress caused stress concentration, resulting in dislocations [4]. This effect became more obvious with the increase in tensile temperature, which led to the weakening of the strain strengthening effect, accelerating the uniform plastic deformation period, as well as reducing the uniform plastic deformation period. As a result, the specimen produced local plastic deformation earlier, that is, the phenomenon of "retraction neck".

The tensile fracture microstructures of the joint specimens at different temperatures are shown in Figure 7. The fracture location of the room temperature tensile joint occurred in the center of the weld, and there was a large area of dimples on the fracture. These dimples were closely distributed and mostly equal axial or slightly elongated dimples (Figure 7a), showing typical ductile fracture characteristics. The high-temperature tensile fracture was mainly composed of dimples, cleavage planes, and slip. In the local part, when the external stress gradually increased, there was a "retraction neck" phenomenon. When it shrank to a certain extent, tearing occurred, resulting in the overall fracture of the weld and forming dimples [17]. Moreover, Figure 7b indicates that the fracture mechanism of the high-temperature tensile specimens was a mixed tough–brittle fracture.

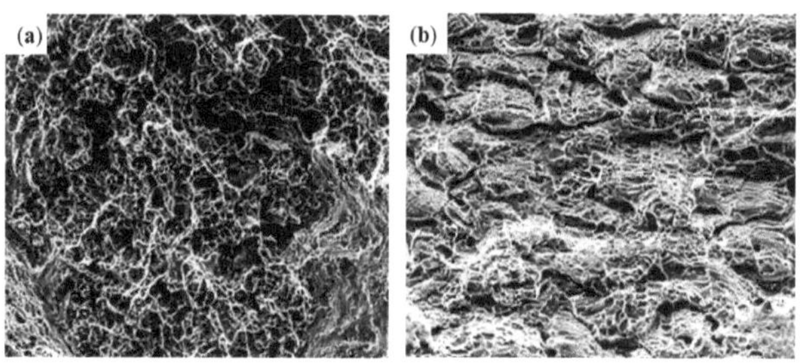

Figure 7. Tensile fracture microscopic morphology: (**a**) room temperature; (**b**) high temperature.

5. Numerical Simulation of High-Temperature Tension of Welded Joints

An axisymmetric finite element model of the welded joint was established for the numerical simulation analysis. The physical parameters of the TP304H/T22 dissimilar steel welded joints used for the numerical simulation are shown in Table 5. Mesh division was performed using CAX4R linear reduced integration elements. The finite element model comprised 3492 nodes and 3302 elements. To simulate the real unidirectional tensile test of the welded joint of the TP304H/T22 dissimilar steel at 550 °C, the temperature field was predefined in the tensile region of the welded joint model. One end of the welded joint model was constrained, and a displacement load was applied to the other end. Figure 8 demonstrates the results of the numerical simulation and the tensile test.

Table 5. Physical properties of the TP304H/T22 welded joint at 550 °C.

Material	Density (t/mm^3)	Expansion ($°C^{-1}$)	Poisson Ratio	Young's Modulus (MPa)	Yield Stress (MPa)	Plastic Strain
T22	7.8×10^{-9}	1.46×10^{-5}	0.3	169,000	105 210	0 0.02
TP304H	7.8×10^{-9}	1.86×10^{-5}	0.31	156,000	159 320	0 0.02
N06600	7.8×10^{-9}	1.5×10^{-5}	0.31	182,000	240 560	0 0.02

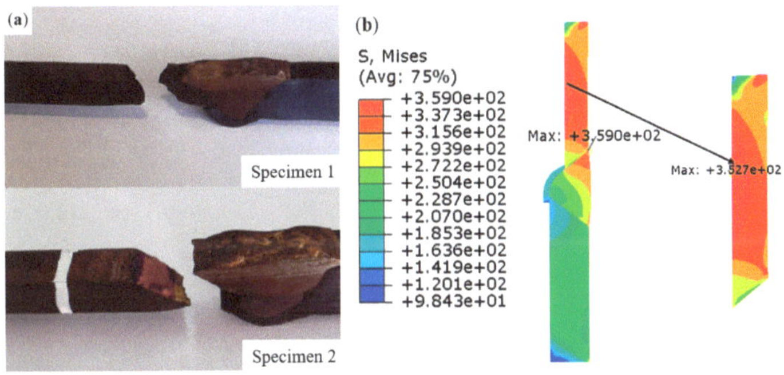

Figure 8. High-temperature tensile testing of the welded joints: (**a**) experiment; (**b**) simulation.

In actual working conditions, the superheater pipeline is mainly subjected to circumferential tension caused by high-temperature and high-pressure steam, as well as bending stress and axial tensile stress caused by secondary stress [2]. The stress concentration effect increases the local stress value, which is the weak position of the joint. The stress distribution of welded joints should be paid more attention. Based on the distribution law of equivalent stress in unidirectional tensile, the danger zone of a welded joint was defined to avoid cracking due to the stress concentration during service. In Figure 8b, two high stress zones can be observed in the TP304H base metal and the TP304H side weld fusion zone of the joint. According to the finite element simulation, the maximum equivalent stress concentrated on the base metal of TP304H at the initial stage of loading, and the effect of stress concentration played a dominant role. In the middle and late stages of loading, because of the different properties of the materials in each section, nonuniform deformation occurred in the tensile process, and the maximum equivalent stress appeared in the weld fusion zone of the TP304H side. The interaction between the different materials at the weld position was more obvious. In the actual damage process, the damage would preferentially appear in the two high stress zones and then evolve into microcracks, which together with the macroscopic crack propagation would lead to the fracture of the specimen, as shown in Figure 8a.

6. Conclusions

In this study, the microscopic morphology, microhardness, and tensile properties of TP304H/T22-type dissimilar steel welded joints in long-term service conditions were analyzed, and the following conclusions were obtained:

(1) No damage features, such as creep cavities and intergranular cracks, were found in the microstructure of the welded joints. On the side of the TP304H close to the fusion line, austenite grains grew from the base metal matrix to the weld recrystallization zone. The microstructure of the heat-affected zone (HAZ) on the T22 side was refined, which provided the HAZ with better crack initiation resistance.

(2) The microhardness test results show that tube sample weld zone and base metal had a significant hardness difference. The microstructure analysis revealed that the strong resistance of the grains to dislocation movement during deformation and the significant increase in dislocation proliferation resulted in the highest hardness values at the weld and that crack initiation and propagation were more likely to occur.
(3) The high-temperature tensile test of the welded joint showed that fracture occurs in the base metal, or the TP304H side fusion zone. In addition, the numerical simulation results also show that in the actual high-temperature tensile process, in those two parts of the high stress area, damage defects were prone to initiation. The high-temperature tensile fracture morphology indicates that the fracture form was a mixed fracture. The joint fractured at the weld under room-temperature tensile tests, and the fracture showed the characteristics of a ductile fracture.
(4) The tensile test results showed that the welded joints fractured at different locations after a long period of service than those unserved. Combined with the numerical simulation results, the weak position of the welded joints of the in-service dissimilar steel was clarified. This is a reference for the safe and stable operation of in-service dissimilar steel welded joints.

Author Contributions: Funding acquisition, H.L. (He Lin) and H.L. (Hui Liu); Project administration, X.Z.; Resources, F.L. and Z.Z.; Methodology, H.L. (He Lin), H.L. (Hui Liu) and X.Z.; Writing—original draft, T.W.; Writing—review and editing, J.S., Y.C. and X.C. All authors have read and agreed to the published version of the manuscript.

Funding: This research was financially supported by the National Natural Science Foundation of China (Program No. 51805402), Natural Science Basic Research in Shaanxi Province of China (Program No. 2022JM-362), QINCHUANGYUAN "Engineer plus Scientist" Team of Shaanxi Province, China (Program No. 2023KXJ-129), Xi'an Key Laboratory of Modern Intelligent Textile Equipment in Shaanxi Province of China (Program No. 2019220614SYS021CG043).

Institutional Review Board Statement: Not applicable.

Informed Consent Statement: Not applicable.

Data Availability Statement: Not applicable.

Conflicts of Interest: The authors declare no conflict of interest.

References

1. Ratković, N.; Lazić, V.; Arsić, D.; Nikolić, R.R.; Prokić, C.R.; Popović, O. Microstructure in the Joining Zone during the Friction Welding of the Two Dissimilar Steels. *Ind. Lubr. Tribol.* **2018**, *70*, 401–407. [CrossRef]
2. Fan, Z.D.; Zhang, Z.B.; Xu, G.; Ma, Y.C.; Niu, K.; Tang, W.S. Failure Reason Analysis of TP304H/T22 Type Dissimilar Steel Weld after Serving for Long Time. *Hot Work. Technol.* **2019**, *48*, 175–178+182.
3. Karthick, K.; Malarvizhi, S.; Balasubramanian, V. Microstructural Characterization of Dissimilar Weld Joint between Ferritic Steel and Stainless Steel. *Mater. Sci. Technol.* **2021**, *37*, 1257–1269. [CrossRef]
4. Ghosh, A.; Sahu, M.; Singh, P.K.; Kumar, S.; Ghosh, M.; Sagar, S.P. Assessment of Mechanical Properties for Dissimilar Metal Welds: A Nondestructive Approach. *J. Mater. Eng. Perform.* **2019**, *28*, 900–907. [CrossRef]
5. Ravikiran, K.; Das, G.; Kumar, S.; Singh, P.K.; Sivaprasad, K.; Ghosh, M. Evaluation of Microstructure at Interfaces of Welded Joint between Low Alloy Steel and Stainless Steel. *Metall. Mater. Trans. A* **2019**, *50*, 2784–2797. [CrossRef]
6. Zhang, X.; Mi, G.; Wang, C. Microstructure and Performance of Hybrid Laser-Arc Welded High-Strength Low Alloy Steel and Austenitic Stainless Steel Dissimilar Joint. *Opt. Laser Technol.* **2020**, *122*, 105878. [CrossRef]
7. Bai, X.L.; Zhang, Q.; Chen, G.H.; Wang, J.Q.; Liu, J.J.; Yu, X.H.; Hua, J.; Zhang, J.H.; Tang, W.M. High Temperature Tensile Test and Creep Rupture Strength Prediction of T92/Super304H Dissimilar Steel Weld Joints. *Mater. High Temp.* **2014**, *31*, 69–75. [CrossRef]
8. Zhang, K.; Ni, L.; Lei, Z.; Chen, Y.; Hu, X. Microstructure and Tensile Properties of Laser Welded Dissimilar Ti-22Al-27Nb and TA15 Joints. *Int. J. Adv. Manuf. Technol.* **2016**, *87*, 1685–1692. [CrossRef]
9. Ou, P.; Cao, Z.; Hai, M.; Qiang, J.; Wang, Y.; Wang, J.; Zheng, G.; Zhang, J. Microstructure and Mechanical Properties of K-TIG Welded Dissimilar Joints between TC4 and TA17 Titanium Alloys. *Mater. Charact.* **2023**, *196*, 112644. [CrossRef]
10. Zhai, X.; Du, J.; Li, L.; Zhou, C.; Zhang, Z. Creep Behavior and Damage Evolution of T92/Super304H Dissimilar Weld Joints. *J. Iron Steel Res. Int.* **2019**, *26*, 751–760. [CrossRef]

11. Mittal, R.; Sidhu, B.S. Microstructural and Mechanical Studies of T91/T22 Welded Joints. *Trans. Indian Inst. Met.* **2018**, *71*, 1985–2000. [CrossRef]
12. Bettahar, K.; Bouabdallah, M.; Badji, R.; Gaceb, M.; Kahloun, C.; Bacroix, B. Microstructure and Mechanical Behavior in Dissimilar 13Cr/2205 Stainless Steel Welded Pipes. *Mater. Des.* **2015**, *85*, 221–229. [CrossRef]
13. Wang, W.; Liu, T.G.; Cao, X.Y.; Lu, Y.H.; Shoji, T. In-Situ SEM Study of Crack Initiation and Propagation Behavior in a Dissimilar Metal Welded Joint. *Mater. Sci. Eng. A* **2018**, *729*, 331–339. [CrossRef]
14. Chen, C.; Li, Q.Q.; Lin, S.B.; Zhou, J.; Fan, C.L.; Yang, C.L. Mechanism of Crack Initiation of Dissimilar Joints be-tween 2219 Aluminum Alloy and 5A06 Aluminum Alloy. *Rare Met. Mater. Eng.* **2019**, *48*, 775–781.
15. Sun, Q.; Liu, Q.; Li, X.; Cai, Z.; Huo, X.; Fan, M.; He, Y.; Yao, X.; Pan, J.; Li, K. Near-Threshold Fatigue Crack Growth Behavior of the Weld Metal and Heat-Affected Zone in an Ex-Service CrMoV Steel Joint. *Int. J. Fatigue* **2022**, *164*, 107175. [CrossRef]
16. Aghajani Derazkola, H.; Khodabakhshi, F. Intermetallic Compounds (IMCs) Formation during Dissimilar Friction-Stir Welding of AA5005 Aluminum Alloy to St-52 Steel: Numerical Modeling and Experimental Study. *Int. J. Adv. Manuf. Technol.* **2019**, *100*, 2401–2422. [CrossRef]
17. Chen, W.; Wang, W.; Liu, Z.; Zhai, X.; Bian, G.; Zhang, T.; Dong, P. Improvement in Tensile Strength of Mg/Al Alloy Dissimilar Friction Stir Welding Joints by Reducing Intermetallic Compounds. *J. Alloys Compd.* **2021**, *861*, 157942. [CrossRef]
18. Zhang, C.; Sun, J.; Li, Y.; Deng, D. Numerical Simulation of Residual Stress of Butt-Welded Joint Involved in Complex Column-Beam Welded Structure. *J. Manuf. Process.* **2022**, *83*, 458–470. [CrossRef]
19. Adomako, N.K.; Shin, G.; Park, N.; Park, K.; Kim, J.H. Laser Dissimilar Welding of CoCrFeMnNi-High Entropy Alloy and Duplex Stainless Steel. *J. Mater. Sci. Technol.* **2021**, *85*, 95–105. [CrossRef]
20. Sun, Y.; Xue, H.; Yang, F.; Wang, S.; Zhang, S.; He, J.; Bashir, R. Mechanical Properties Evaluation and Crack Propagation Behavior in Dissimilar Metal Welded Joints of 304 L Austenitic Stainless Steel and SA508 Low-Alloy Steel. *Sci. Technol. Nucl. Install.* **2022**, *2022*, e3038397. [CrossRef]
21. Wang, Z.T.; Jiang, S.Y.; Chen, Y.P.; Xiong, Z.Y.; Hu, D.A.; Cheng, D.H. Numerical simulation of laser welding of magnesium and steel dissimilar metal based on ABAOUS. *Trans. Mater. Heat Treat.* **2021**, *42*, 134–141.
22. Huang, B.; Fang, Z.; Yang, J.; Zheng, J.; Wang, S. Numerical Simulation of S355JR-316L Dissimilar Metal Welding. *Weld. World* **2022**, *66*, 287–299. [CrossRef]
23. Danielewski, H.; Skrzypczyk, A.; Tofil, S.; Witkowski, G.; Rutkowski, S. Numerical Simulation of Laser Welding Dissimilar Low Carbon and Austenitic Steel Joint. *Open Eng.* **2020**, *10*, 491–498. [CrossRef]
24. Perulli, P.; Dassisti, M.; Casalino, G. Thermo-Mechanical Simulation of Hybrid Welding of DP/AISI 316 and TWIP/AISI 316 Dissimilar Weld. *Materials* **2020**, *13*, 2088. [CrossRef] [PubMed]
25. Pazooki, A.M.A.; Hermans, M.J.M.; Richardson, I.M. Finite Element Simulation and Experimental Investigation of Thermal Tensioning during Welding of DP600 Steel. *Sci. Technol. Weld. Join.* **2017**, *22*, 7–21. [CrossRef]
26. Schmidová, E.; Sunil Kumar, M.R.; Schmid, M.; Bozkurt, F. Role of Nb in the Failure of Dual-Phase Steel in Heterogeneous Welds. *Eng. Fail. Anal.* **2020**, *116*, 104708. [CrossRef]
27. Kumar, R. Behaviours of Oxides Scale of Welded T22 Steels in Thermal Cyclic Conditions. *Mater. Lett.* **2022**, *308*, 131215. [CrossRef]
28. Okonkwo, B.O.; Ming, H.; Wang, J.; Meng, F.; Xu, X.; Han, E.-H. Microstructural Characterization of Low Alloy Steel A508–309/308L Stainless Steel Dissimilar Weld Metals. *Int. J. Press. Vessels Pip.* **2021**, *190*, 104297. [CrossRef]

materials

MDPI

Article

Assessment of Changes in Abrasive Wear Resistance of a Welded Joint of Low-Alloy Martensitic Steel Using Microabrasion Test

Krzysztof Ligier *, Jerzy Napiórkowski and Magdalena Lemecha

Faculty of Technical Sciences, University of Warmia and Mazury in Olsztyn, Ul. M. Oczapowskiego 11, 10-719 Olsztyn, Poland; jerzy.napiorkowski@uwm.edu.pl (J.N.); magdalena.lemecha@uwm.edu.pl (M.L.)
* Correspondence: krzysztof.ligier@uwm.edu.pl

Abstract: Martensitic low-alloy steels are widely used in machine construction. Due to their declared weldability, arc welding is most often used to join elements made of this type of steel. However, the high temperature associated with welding causes unfavourable changes in the microstructure, resulting in reduced abrasion resistance. Therefore, it is important to know the tribological properties of the welded joint. This article presents the results of a study on the abrasion wear resistance of a welded joint of an abrasion-resistant steel. This study tested a welded joint of an abrasive-resistant steel produced by the arc welding method. Wear testing of the welded joint was carried out under laboratory conditions by the ball-cratering method in the presence of abrasive slurry on the cross-section of the welded joint. Based on the test results, the change in the abrasive wear rate of the material as a function of the distance from the welded joint axis was determined. It was also found that the thermal processes accompanying welding caused structural changes that increased the wear rate index value. Adverse changes in the tribological properties of a welded material persist up to a distance of approx. 20 mm from the weld centre.

Keywords: wear testing; welded joint; abrasive wear; ball-cratering method

check for updates

Citation: Ligier, K.; Napiórkowski, J.; Lemecha, M. Assessment of Changes in Abrasive Wear Resistance of a Welded Joint of Low-Alloy Martensitic Steel Using Microabrasion Test. *Materials* **2024**, 17, 2101. https://doi.org/10.3390/ma17092101

Academic Editors: Swarup Bag and Christ Prakash Paul

Received: 13 March 2024
Revised: 16 April 2024
Accepted: 23 April 2024
Published: 29 April 2024

1. Introduction

Low-alloy martensitic steels, thanks to their workability and weldability as declared by their manufacturers and their resistance to abrasive wear, are widely used in the mining, agricultural, and transport industries for operating parts exposed to abrasive wear [1,2]. Due to the aforementioned functional properties of this steel type, machinery users often use these materials to shape machine operating parts from them, according to their needs, while usually using welding techniques. However, despite the declared weldability of these steel, adverse changes in the microstructure, caused by welding, are observed in the welding area [3–7], resulting in a reduction in the abrasion resistance commonly associated with hardness [8–11]. A change in the hardness of high-strength materials under the influence of welding is a widely analysed issue in the literature [3,12–16]. Previous studies [14,15,17] demonstrated that welding processes adversely affected resistance to abrasive wear within the weld metal material zone and the heat-affected zone through structural changes in these areas. The analysis of changes in the microstructure of materials under the influence of welding processes was carried out in detail [5,12–14,16,17] and concerns individual areas of the joint, from the base material (BM), through the heat-affected zone (HAZ) to various joint zones. Despite the accurate testing of the microstructure of individual zones of a welded joint, their tribological properties are no longer studied in such detail. Under laboratory conditions, methods of the type "rubber wheel-dry abrasive", based on standards ASTM G65 [18] and GOST 23.208-79 [19], are commonly used to assess the wear resistance of welded joints [15–17]. Since the size of the test area using these methods can amount to several cm^2, the testing covers multiple zones of the welded joint at the same time.

Therefore, the abrasive wear resistance determined by these methods is a result of the wear resistance of individual areas of the joint. If an accurate assessment of the tribological properties of individual zones of a welded joint is required, methods should be used in which the area covered by a single test falls within one tested joint zone. A method for testing the abrasive wear rate, which satisfies this requirement, is the ball-cratering method.

The ball-cratering method is an abrasive wear test method which is widely used for the assessment of the abrasive wear rate of a wide range of construction materials, e.g., metals [20], ceramics [21], polymers [22,23], and thin coatings [24]. A characteristic property of the ball-cratering method is the formation of small wear marks, which are used to determine the abrasive wear rate for a material. This property of the ball-cratering method can be used to analyse the abrasive wear resistance of individual zones in the cross-section of a welded joint [25].

The aim of this work is to assess the suitability of the ball-cratering method for the analysis of wear resistance of a welded joint in its cross-section. This study aimed to determine changes in abrasive wear resistance of the cross-section of a welded joint of low-alloy martensitic steel, depending on the distance from the weld centre.

2. Materials and Methods

2.1. Welding Process

The joint was prepared using one of the commercially available low-alloy martensitic steels (ESTI s.r.l., Idro, Italy). The chemical composition, as declared by the manufacturer, and the strength properties of the test steel are provided in Table 1.

Table 1. Declared chemical composition and strength properties of test steel.

Chemical Element	C	Si	Mn	P	S	Cr	Ni	Mo	B
Content [%]	0.28	0.35	1.40	Max. 0.30	Max. 0.03	0.50	0.30	0.25	Max. 0.004
Declared hardness	HB 470–530 (over the entire profile thickness after heat treatment)								
Tensile strength Rm	1770 MPa								
Yield point Re	1330 MPa								

The 400 mm long sections were cut off from a ready-made ploughshare heat-treated by the manufacturer, with dimensions of 0.11 m × 0.08 m and a thickness of 12 mm, made from a low-alloy martensitic steel. The sections were then welded along the long side using the butt weld. The process of solid wire welding in active gas shielding (MAG 135) in the downhand position (PA) was used. A semi-automatic welding machine (KEMPPI Fastmig KMS 400 A); 1.2 mm diameter Lincoln Electric SupraMig HD wire [26] (EN ISO 14341:2020) designed for welding in the manufacture of earthmoving, agricultural, and mining machinery (typical chemical composition 0.08% C, 1.70% Mn, 0.85% Si); and C1 shielding gas (100% CO_2) were used.

The following welding parameters were used:

- Welding current I = 230 A;
- Welding voltage U = 30 V;
- Welding speed for manual process about v = 0.35 m/min.

The heat introduced into the joint was calculated from the following formula [27]:

$$Q = \frac{k \cdot U \cdot I \cdot 60}{v \cdot 1000}\left[\frac{\text{kJ}}{\text{mm}}\right] \quad (1)$$

where Q is the line energy [kJ/mm]; U is the voltage (U = 30 V); I is the current (I = 230 A); k is the thermal efficiency coefficient of the welding process (k = 0.8); and v is the welding speed (v = 350 mm/min).

The calculated heat input was equal to 0.946 kJ/mm. The manufacturer of the sheet to be welded allows sheet thicknesses up to 12 mm to be welded without preheating. Due to possible fluctuations in the welding speed, the heat input may exceed 1 kJ/mm. According to the sheet metal manufacturer's recommendations, if the heat input is between 1.0 and 1.69 kJ/mm, it is recommended to increase the preheating temperature by 25 °C above the recommended preheating temperature. It was therefore decided to preheat the joined sheets to 50 °C. The preheating process was carried out in a muffle furnace for 60 min.

The sealing run (1) was made first, followed by fill-up passes (2, 3) and one capping (4), forming the face of the joint, made in subsequent stages of the welding process. The application of interpass cooling to 225 °C was carried out in air. The temperature was controlled using a pyrometer. The produced joint was not heat-treated.

2.2. Sample Preparation

The produced welded joint was cut transversely into 10 mm wide sections from which specimens were prepared for metallographic testing as well as hardness and wear rate testing. The testing of the hardness and wear rate of the produced joint was carried out at a depth of 3 mm from the top edge of the sample (L1 line—Figure 1).

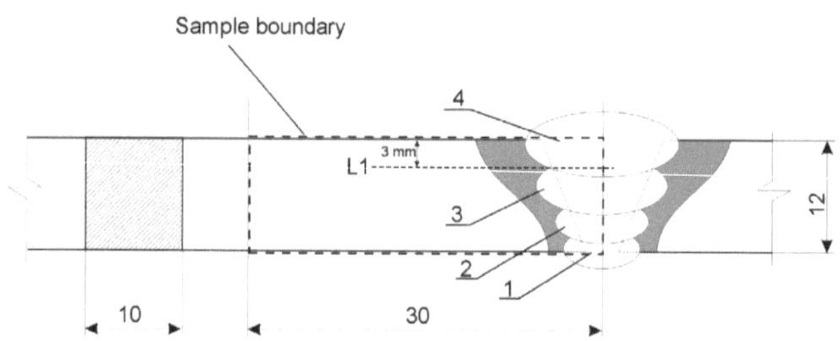

Figure 1. Diagram of the welded joint under test: 1—sealing run, 2, 3—fill-up passes, 4—capping, L1—hardness and wear rate test line.

The transverse surfaces were ground and polished using the Struers LaboPol-5 (Struers S.A.S., Champigny-sur-Marne, France) polisher, and the specimens for metallographic testing were additionally etched with nital (a 5% HNO_3 alcoholic solution).

The assessment of the joint microstructure was carried out by light microscopy methods using a Keyence VHX 700 (Mechelen, Belgium) digital optical microscope.

The hardness of the welded joint was determined by the Vickers method in accordance with standard ISO 6507-1:2018 [28] using a Wilson VH1150 (London, UK) with a load of 98.1 N and a load application time of 10 s. Hardness measurements were taken every 1 mm along the L1 line (Figure 1). The hardness assessment was carried out for five specimens.

Abrasive wear rate testing was carried out by the ball-cratering method using a tribometer with a fixed ball system (Łukasiewicz Institute for Sustainable Technologies, Radom, Poland) (Figure 2).

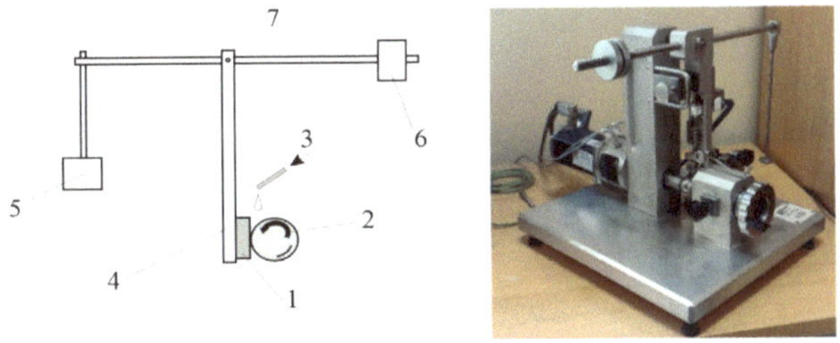

Figure 2. Schematic and general views of the test stand for testing the abrasive wear using the ball-cratering method. 1—sample, 2—ball (counter-sample), 3—abrasive slurry feed, 4—sample holder arm, 5—load, 6—counterweight, 7—pivot.

A 25.4 mm (1″) diameter ball made of 100Cr6 steel, with a hardness of 58.6 HRC, was used as the counter-specimen. The abrasive wear testing was carried out according to standard EN-1071-6:2008 [27,29], using the following parameters:

- Friction assembly load: 0.4 N;
- Counter-specimen rotational speed: 150 rpm;
- Experimental run duration: 15 min;
- Sliding distance: 179.5 m.

The abrasive wear testing was carried out using an abrasive slurry prepared from aluminium oxide (Al_2O_3) with a grain size of 3 μm (P.P.U.H. "KOS" Stanisław Kos, Koło, Poland) and distilled water. The slurry volume concentration was approx. 2%. The slurry was fed onto the friction assembly in an amount of 1 cm^3/min. The ball and the test surface were washed and degreased with ethyl alcohol each time.

Due to the limitations of the tribometer's working area, the tribological test specimens covered the welded joint from the weld axis to the base material. Test runs were made every 2 mm along the line perpendicular to the joint axis, located 3 mm from the top edge of the sheet being joined (L1 line in Figure 1), with a test area length of 30 mm. The testing was carried out in six replications using six specimens.

The wear volume was determined based on the diameter of the obtained craters, measured perpendicular and parallel to the direction of movement of the abrasive particles. The crater diameters were measured using a digital optical microscope.

The wear volume was calculated using the following formula:

$$V = \frac{\pi \cdot b^4}{64 \cdot R}\left[\mathrm{mm}^3\right] \tag{2}$$

where R is the ball radius [mm] and b is mean crater diameter [mm].

The wear rate index was calculated using the Archard formula [29] (EN-1071-6:2008):

$$W_r = \frac{V}{S \cdot N} = \frac{\pi b^4}{64 \cdot R \cdot S \cdot N}\left[\mathrm{mm}^3 \cdot \mathrm{N}^{-1} \cdot \mathrm{m}^{-1}\right] \tag{3}$$

where W_r is the wear rate index [mm^3 N^{-1} m^{-1}], R is the ball radius [mm], b is the mean crater diameter [mm], S is the friction distance [m], and N is the friction assembly normal load [N].

3. Results and Discussion

3.1. Joint Microstructure

Several impacts with a concentrated heat source in the welding process resulted in noticeable changes in the microstructure in both the base material and the welded joint. In the test area, five zones with different microstructure morphologies were distinguished. These zones are marked in Figure 3.

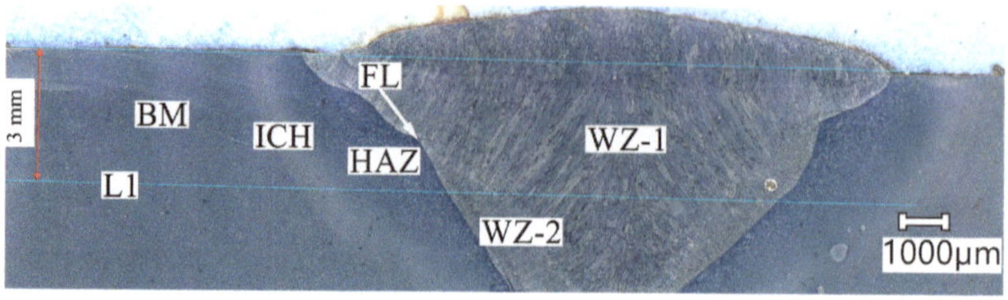

Figure 3. A view of the welded joint section under study: BM—base material; ICH—intercritical zone; HAZ—heat-affected zone; WZ-1 and WZ-2—weld zones; FL—fusion line; L1—testing line.

The base material (BM) (Figure 4a) exhibited a tempered martensite © microstructure typical of low-alloy boron steels.

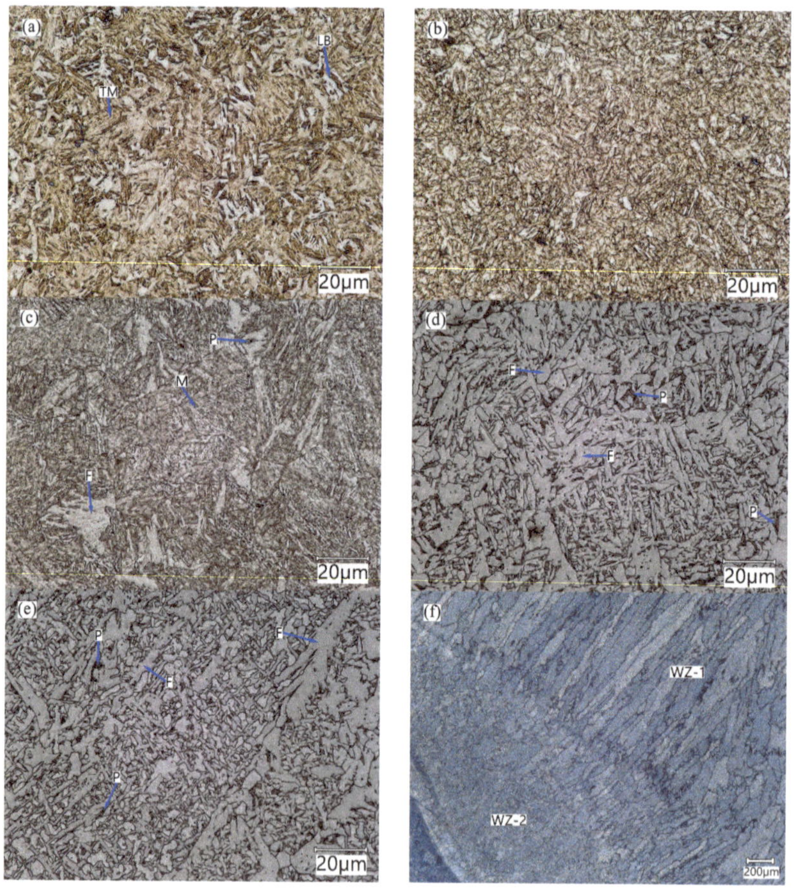

Figure 4. The microstructure of distinguished areas of the welded joint: (**a**) base material, (**b**) partially heat-affected zone (PHAZ), (**c**) heat-affected zone (HAZ), (**d**) weld zone within the fusion penetration zone (WZ-2), (**e**) weld filling (WZ-1), (**f**) visible differences in the microstructure orientation in zones WZ-1 and WZ-2; (TM—tempered martensite; LB—lower bainite; P—perlite; M—martensite; F—ferrite).

Occasionally, individual grains of lower bainite (LB) can also be discerned. The heating process during welding and the rapid heat removal resulted in the transformations of the martensite microstructure (Figure 4b) in the ICH area. The intercritical heat-affected zone (ICHAZ) experiences a peak temperature between Ac1 and Ac3 and has a mixed structure of fine re-austenitized grains and tempered martensite retained from the base metal [30–32]. In the intercritical zone (ICH), the welded joint is characterised by the presence of tempering martensite and troostite. The observed type of structure of the intercritical zone of the welded joint is similar to that found in low-carbon low-alloy steels subjected to underhardening quenching [33]. In the HAZ microstructure, areas of tempered martensite and ferrite along with small areas of perlite were observed (Figure 4c).

The joint area, including the fill-up passes, is ferrite of varying morphology and dispersion. Within the weld zone WZ-1, structures typical of varying temperatures and cooling rates can be observed. The microstructure of this area is formed by acicular ferrite with perlite areas (Figure 4d). In the WZ-2 zone (Figure 4e), which includes the recrystallised area of the weld, acicular ferrite (F) is mainly found. In the WZ-2 area, as a result of thermal effects caused by successive transitions of the welding process, a split of the columnar microstructure into a (nearly) equiaxial (fragmented) microstructure was observed, which is typical of the annealing that occurs during multi-run welding [34]. The fragmentation of the ferritic structure results from multiple recrystallisation of the phase components of the joint and rapid heat removal. It should be noted that the clear orientation of the microstructure towards rapid heat removal is noticeable in the WZ-1 zone (Figure 4f).

3.2. *Joint Hardness*

The obtained results (Figure 5) indicate a change in the mechanical properties of the welded material, depending on the distance from the weld centre.

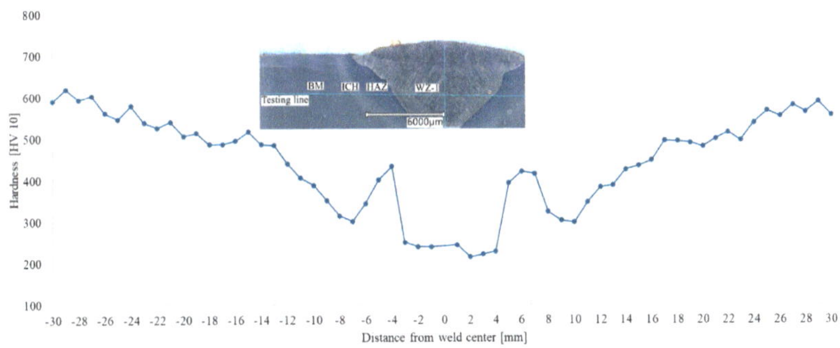

Figure 5. Hardness profile of one of the tested welded joint specimens.

Within the weld area, the lowest hardness values (230 HV10) were noted in relation to the material not subjected to heat impact. The lowest hardness values were noted in WZ-1 and WZ-2 zones, which have values approx. 2.5 times lower than the hardness of the base material (approx. 600 HV 10). This should be associated with the fact that the WZ-1 and WZ-2 zones consist mainly of ferrite phases with different morphologies and small areas of perlite. The local increase in hardness near the fusion line and its subsequent decrease in the HAZ are linked to the changes in the material microstructure in this area of the joint, resulting from the welding process [34].

An increase in hardness near the fusion line results from the supercritical temperature of this area reached in the welding process and the rapid heat removal. In the heat-affected zone, pearlite and tempered martensite appear, resulting in an increase in hardness. The decrease in the hardness of the material located further away from the fusion line is due to the lower heating temperature of this area resulting in the further tempering of the martensite. This phenomenon occurs in welded joints of low-alloy steel not subjected to subsequent heat treatment [3,12,14,35].

3.3. Abrasive Wear Resistance

Figure 6 shows examples of craters obtained within different zones of the test welded joint.

Figure 7 shows the obtained results for the abrasive wear rate index for the welded joint, depending on the distance from the weld axis.

It can be noted that different values of the abrasive wear rate index were noted for the distinguished areas of the test welded joint. The highest abrasive wear rate occurs in the WZ-1 area located immediately near the weld axis, which is due to the low hardness of this area, resulting from the greatest proportion of welding material in the weld material as well as the greatest changes in the microstructure [34]. The WZ-2 zone, characterised by a microstructural structure similar to that of the WZ-1 zone but with grains oriented to a lesser extent, exhibits an abrasive wear rate lower than that for WZ-1. A decrease in the wear rate can be seen approx. 5 mm from the weld axis. This area is the beginning of the heat-affected zone where there is no impact of the additional material, and the wear resistance is determined by the microstructure. Due to the presence of martensite in the microstructure, the zone exhibits the greatest hardness and, thus, a lower wear rate than that in the weld area [25,36]. An increase in the abrasive wear rate is then observed at a distance of 7–9 mm from the weld centre, followed by its systematic decrease as the distance from the joint axis increases.

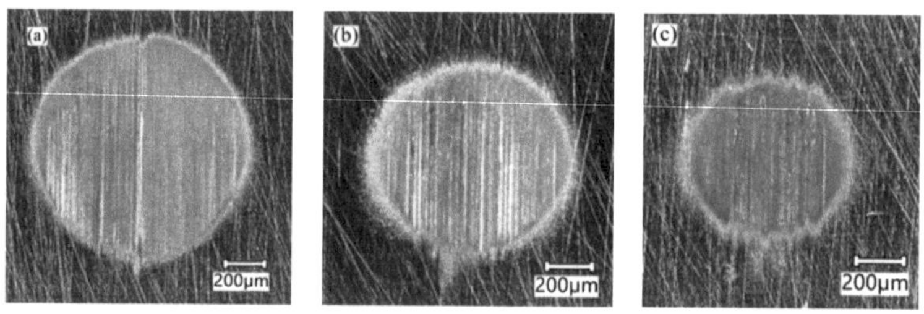

Figure 6. Examples of craters obtained within different zones of the test welded joint: (**a**) 1 mm distance from the joint axis—(WZ-1); (**b**) 5 mm distance from the joint axis—(HAZ); (**c**) 23 mm distance from the joint axis—(BM).

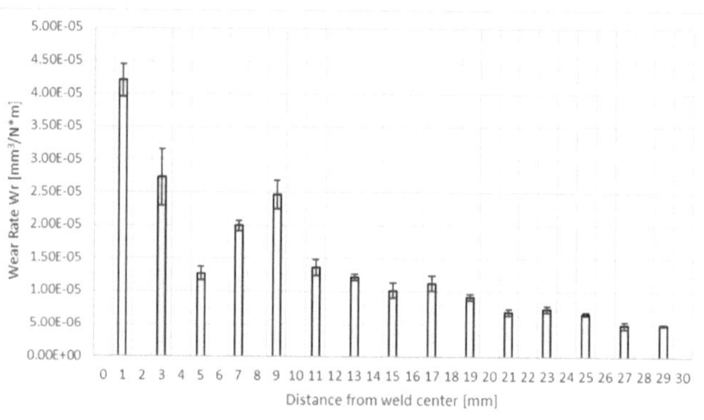

Figure 7. Average wear rate index values for the test welded joint, depending on the distance from the weld centre. Error bars—standard deviation.

When comparing the course of changes in the wear rate index and the hardness in the cross-section of the welded joint as a function of distance from the weld centre, it can be noted that there is a correlation between these quantities (Figure 8).

The increase in hardness corresponds to the reduction in wear rate, which is in line with the common view in the literature, resulting from the Archard equation that links abrasive wear resistance to the hardness of the material, which is widely presented in the literature [30–32,37–39]. The variations in the wear rate are due to the different heating temperatures of the base material during the welding process. The closer to the joint axis, the higher the tempering temperature of the martensite, which is consistent with the literature [36,40]. The authors [40] report that there is a relation between the microstructural transformation as well as the mechanical properties and the tribological response of low-alloy wear-resistant martensitic steel during sliding wear.

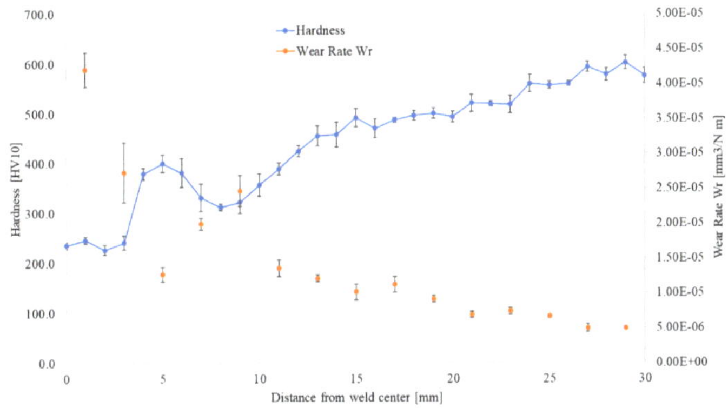

Figure 8. A comparison of the distribution of hardness and wear rate depending on the distance from the weld centre. Error bars—standard deviation.

It is worth noting that the observed changes in the microstructure in the test area extend to approx. 10 mm (heat-affected zone boundary), while the abrasive wear rate being more than twice as high as that for the base material was noted at a distance of 17 mm from the weld centre. This indicates that the change in the tribological properties is already noticeable despite the lack of observed changes in the microstructure.

Large fluctuations in the obtained values of the wear rate and hardness index were also observed. This can be related to the accuracy of the positioning of the samples in the tribometer and the manual welding process. The accuracy of the specimen positioning overlapped with the variable weld width. Wear tests were carried out on untreated surfaces and specimen alignment was based only on the distance from the weld joint axis. The test specimens had a thickness of 10 mm, and such a distance separates the different tested joint surfaces. Due to the manual welding process, the width of the weld being produced can vary, and consequently, the boundaries of individual zones in the cross-section can be located at different distances from the weld centre. This has resulted in the fact that some test runs (despite efforts) may have been conducted in different sub-zones of the joint under test.

Three different modes of abrasive wear can be distinguished, i.e., cutting, ploughing, and wedge formation [41]. In all these modes, grooves are formed on the worn surface. However, only the cutting mode leads to material removal, i.e., wear, while the ploughing and wedge formation modes mainly lead to plastic deformation of the surface material. Therefore, the resulting abrasive wear factor will depend on the dominant wear mode in the actual abrasive contact. A high abrasive wear rate is the result of a combination of high hardness, which reduces the penetration rate of the abrasive grains, and high fracture and peel strength of the material [42].

In all the tested welded joint zones, the dominant wear process is ploughing, caused by the action of sharp-edged abrasive grains on the material surface. Ploughing is accompanied by plastic deformation with an intensity that depends on the microstructure of the material in the tested zone.

On the surface of the crater made in the area of the base material (approx. 23 mm from the weld axis), scratches and grooves run parallel to the direction of abrasive travel along the surface (Figure 9). The grooves are narrow and shallow, and there is only occasional slight plastic deformation of the material in the form of indentations and material spreading in their area. The cutting of irregularities is the main abrasive wear type in this case [25,36].

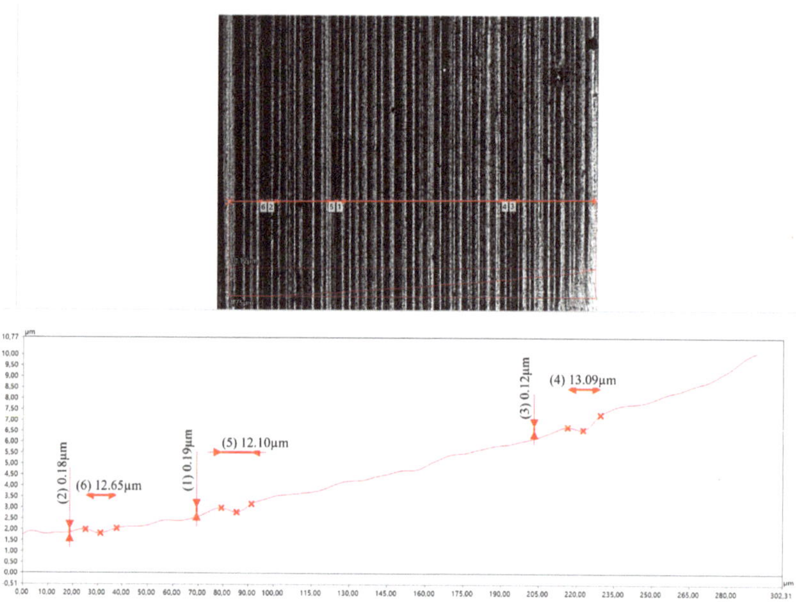

Figure 9. The surface of the crater made 23 mm from the axis of the welded joint.

As the test area approaches closer (remaining outside the HAZ) to the axis of the welded joint, the grooves become deeper. However, they do not lose their orientation in relation to the direction in which the abrasive moves across the surface of the specimen. The plastic deformation of the material is slightly increased due to the decreasing hardness value of the material [36] (Figure 10).

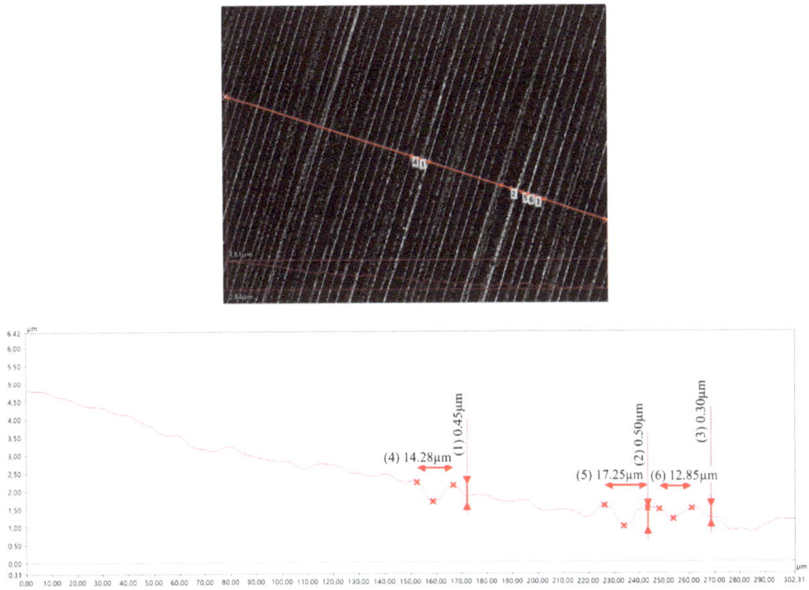

Figure 10. The surface of the crater made 12 mm from the axis of the welded joint.

Approximately 5 mm from the axis of the welded joint, in the heat-affected zone (HAZ), the depth of cracks and grooves decreases slightly (Figure 11). This can be explained by the increased hardness of the microstructure. Furthermore, plastic deformation (pits) is visible in the grooves, which is related to the presence of areas of ferrite in the microstructure [36].

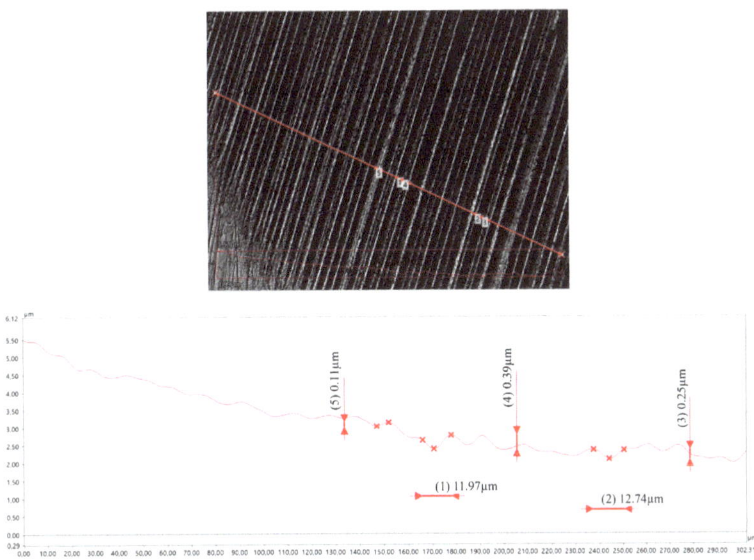

Figure 11. The surface of the crater made 5 mm from the axis of the welded joint.

In the WZ-1 and WZ-2 zones (0 to 4 mm from the welded joint axis), ploughing is still the dominant wear process (Figure 12).

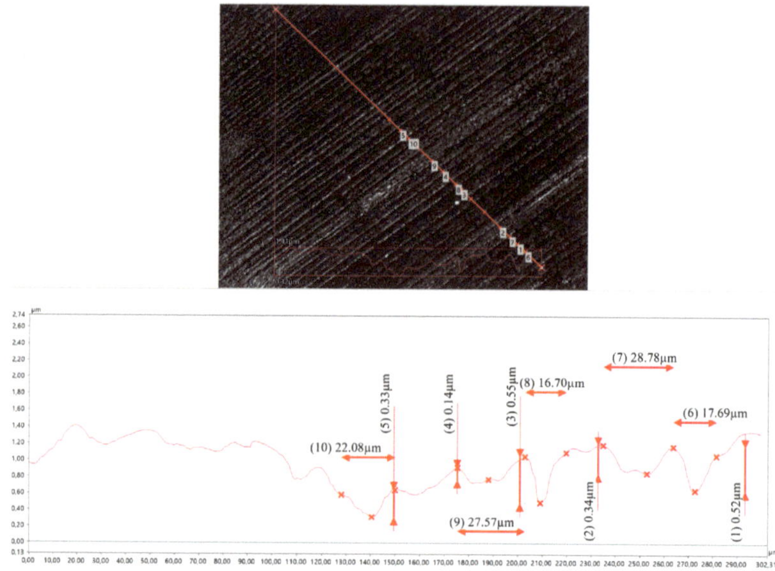

Figure 12. The surface of the crater made 3 mm from the axis of the welded joint.

The scratches and grooves on the specimen surface are characterised by varying depth and width. This is due to the different abrasive wear resistance of ferrite and pearlite, and thus the different abrasive wear mechanisms taking place. The perlite bands are subjected to wear by the shearing of irregularities, while the ferrite is mainly subjected to wear by furrowing with local plastic deformation of the material. The furrows are deep and pushed material is visible at their edges. The plastic deformation takes the form of sharp-edged pits and is more severe than in the HAZ. This is due to the pressing of hard abrasive grains into the relatively soft material [36].

Analysis of the surfaces of the test specimens showed that the softer microstructures are subjected to abrasive wear mainly due to furrowing. In the case of harder and more brittle structures, material loss is caused by micro-cutting, which further smooths the worn surface [36].

Welded joints of low-carbon martensitic steels with boron micro-additives, which are the subject of this study, should be heat-treated. Heat treatment of a welded joint can prolong the life of machine components exposed to abrasive wear.

4. Conclusions

This study's results show that the thermal processes accompanying welding reduce the hardness of the material and increase the wear rate index value. These changes mostly affect the weld area and the heat-affected zone. The weld area of the welded joint proved to be the least resistant to abrasive wear. The wear rate index for this area proved to be greater by 5–8.5 times than that for the base material, with higher values noted closest to the weld centre. The heat-affected zone (HAZ) exhibits a variable abrasive wear rate. In the HAZ area located closest to the weld, a sudden decrease in the wear rate to a value 2.5 times higher than that for the base material was observed, followed by its subsequent increase to a value five times higher (approx. 9 mm from the weld centre) than that for the base material. Adverse changes in the tribological properties of a welded material persist up to a distance of approx. 20 mm from the weld centre.

Due to the relatively small wear marks created by the ball-cratering method, it can be applied to accurately assess the abrasion resistance of selected areas of a welded joint in its cross-section.

Author Contributions: Conceptualization, K.L.; methodology, K.L. and M.L.; validation, K.L., J.N. and M.L.; formal analysis, K.L. and J.N.; investigation, K.L. and M.L.; resources, K.L.; data curation, K.L. and M.L.; writing—original draft preparation, K.L. and M.L.; writing—review and editing, J.N.; visualization, K.L. and M.L.; supervision, K.L. All authors have read and agreed to the published version of the manuscript.

Funding: This research received no external funding.

Data Availability Statement: Data are contained within the article.

Conflicts of Interest: The authors declare no conflict of interest.

References

1. Konat, Ł.; Jasiński, R.; Białobrzeska, B.; Szczepański, Ł. Analysis of the static and dynamic properties of wear-resistant Hardox 600 steel in the context of its application in working elements. *Mater. Sci.-Pol.* **2021**, *39*, 86–102. [CrossRef]
2. Białobrzeska, B.; Jasiński, R.; Konat, Ł.; Szczepański, Ł. Analysis of the Properties of Hardox Extreme Steel and Possibilities of Its Applications in Machinery. *Metals* **2021**, *11*, 162. [CrossRef]
3. Bramowicz, M.; Kulesza, S.; Lewalski, P.; Szatkowski, J. Structural Studies of Welds in Wear-Resistant Steels. *Acta Phys. Pol. A* **2016**, *130*, 4. [CrossRef]
4. Gáspár, M. Effect of Welding Heat Input on Simulated HAZ Areas in S960QL High Strength Steel. *Metals* **2019**, *9*, 1226. [CrossRef]
5. Lu, Y.; Peer, A.; Abke, T.; Kimchi, M.; Zhang, W. Subcritical heat affected zone softening in hot-stamped boron steel during resistance spot welding. *Mater. Des.* **2018**, *155*, 170–184. [CrossRef]
6. Saxena, A.; Kumaraswamy, A.; Madhu, V.; Madhusudhan, R.G. Study of Tribological Characteristics of Multi-pass SMAW Armox 500T Steel Joints. *J. Mater. Eng. Perform.* **2018**, *27*, 4300–4307. [CrossRef]
7. Montero, J.; García, A.; Varela, A.; Zaragoza, S.; Artiaga, R.; Mier, J.L. A study on wear of welded joins for pipelines. *Weld. Int.* **2010**, *24*, 120–124. [CrossRef]
8. Konat, Ł.; Pękalski, G. "Overview of Materials Testing of Brown-Coal Mining Machines" (Years 1985–2017). In *Mining Machines and Earth-Moving Equipment: Problems of Design, Research and Maintenance*; Springer International Publishing: Cham, Switzerland, 2020; pp. 21–58. [CrossRef]
9. Adamiak, M.; Górka, J.; Kik, T. Comparison of abrasion resistance of selected constructional materials. *J. Achiev. Mater. Manuf. Eng.* **2009**, *37*, 375–380.
10. Bhakat, A.K.; Mishra, A.K.; Mishra, N.S. Characterization of wear and metallurgical properties for development of agricultural grade steel suitable in specific soil conditions. *Wear* **2007**, *263*, 228–233. [CrossRef]
11. Białobrzeska, B. The influence of boron on the resistance to abrasion of quenched low-alloy steels. *Wear* **2022**, *500–501*, 204345. [CrossRef]
12. Konat, Ł.; Białobrzeska, B.; Białek, P. Effect of Welding Process on Microstructural and Mechanical Characteristics of Hardox 600 Steel. *Metals* **2017**, *7*, 349. [CrossRef]
13. Brykov, M.N.; Petryshynets, I.; Džupon, M.; Kalinin, Y.A.; Efremenko, V.G.; Makarenko, N.A.; Pimenov, D.Y.; Kovác, F. Microstructure and Properties of Heat Affected Zone in High-Carbon Steel after Welding with Fast Cooling in Water. *Materials* **2020**, *13*, 5059. [CrossRef]
14. Wan, Z.; Guo, W.; Jia, Q.; Xu, L.; Peng, P. Hardness Evolution and High Temperature Mechanical Properties of Laser Welded DP980 Steel Joints. High Temp. *Mater. Proc.* **2018**, *37*, 587–595. [CrossRef]
15. Górka, J. The assessment of the quality of welded joints made of abrasion-resistant plates using the nanocrystalline filler metal. *J. Min. Metall. Sect. B Metall.* **2020**, *56*, 209–220. [CrossRef]
16. Konat, Ł.; Białobrzeska, B. Effect of Welding Technique and Thermal Heatment Parameters on Abrasive Wear of Steel S355. *Tribologia* **2022**, *3*, 25–38. [CrossRef]
17. Tomków, J.; Czupryński, A.; Fydrych, D. The Abrasive Wear Resistance of Coatings Manufactured on High-Strength Low-Alloy (HSLA) Offshore Steel in Wet Welding Conditions. *Coatings* **2020**, *10*, 219. [CrossRef]
18. *ASTM G65*; Standard Test Method for Measuring Abrasion Using the Dry Sand/Rubber Wheel Apparatus. ASTM: West Conshohocken, PA, USA, 2021.
19. *GOST 23.208-79*; Ensuring of Wear Resistance of Products. Wear Resistance Testing of Materials by Friction against Loosely Fixed Abrasive Particles. GostPerevod: Moscow, Russia, 1979.
20. Marques, F.; da Silva, W.M.; Pardal, J.M.; Tavares, S.S.M.; Scandian, C. Influence of Heat Treatments on the Micro-Abrasion Wear Resistance of a Superduplex Stainless Steel. *Wear* **2011**, *271*, 1288–1294. [CrossRef]
21. Antunes, P.V.; Ramalho, A. Study of Abrasive Resistance of Composites for Dental Restoration by Ball-Cratering. *Wear* **2003**, *255*, 990–998. [CrossRef]

22. Farfán-Cabrera, L.I.; Gallardo-Hernández, E.A.; de la Rosa, C.S.; Vite-Torres, M. Micro-Scale Abrasive Wear of Some Sealing Elastomers. *Wear* **2017**, *376–377*, 1347–1355. [CrossRef]
23. Ligier, K.; Olejniczak, K.; Napiórkowski, J. Wear of polyethylene and polyurethane elastomers used for components working in natural abrasive environments. *Polym. Test.* **2021**, *100*, 107247. [CrossRef]
24. Silva, F.; Martinho, R.; Baptista, A. Characterization of Laboratory and Industrial CrN/CrCN/Diamond-Like Carbon Coatings. *Thin Solid Film.* **2014**, *550*, 278–284. [CrossRef]
25. Ligier, K.; Bramowicz, M.; Kulesza, S.; Lemecha, M.; Pszczółkowski, B. Use of the Ball-Cratering Method to Assess the Wear Resistance of a Welded Joint of XAR400 Steel. *Materials* **2023**, *16*, 4523. [CrossRef] [PubMed]
26. *EN ISO 14341:2020*; Welding Consumables Wire Electrodes and Weld Deposits for GAS shielded Metal Arc Welding of Non Alloy and Fine Grain Steels. Classification. International Organization for Standardization: Geneva, Switzerland, 2020.
27. *ISO 15614-1:2017*; Specification and Qualification of Welding Procedures for Metallic Materials Welding Procedure Test. International Organization for Standardization: Geneva, Switzerland, 2017.
28. *ISO 6507-1:2018*; Metallic Materials Vickers Hardness Test. International Organization for Standardization: Geneva, Switzerland, 2018.
29. *EN-1071-6:2008*; Advanced Technical Ceramics—Methods of Test for Ceramic Coatings—Part 6: Determination of the Abrasion Resistance of Coatings by a Micro-Abrasion Wear Test. CEN—European Committee for Standardization: Brussels, Belgium, 2008.
30. Francis, J.A.; Mazur, W.; Bhadeshia, H.K.D.H. Review type IV cracking in ferritic power plant steels. *Mater. Sci. Technol.* **2006**, *22*, 1387–1395. [CrossRef]
31. David, S.A.; Siefert, J.A.; Feng, Z. Welding and weldability of candidate ferritic alloys for future advanced ultrasupercritical fossil power plants. *Sci. Technol. Weld. Join.* **2013**, *18*, 631–651. [CrossRef]
32. Wang, Y.; Kannan, R.; Li, L. Correlation between intercritical heat-affected zone and type IV creep damage zone in grade 91 steel. *Metall. Mater. Trans. A* **2018**, *49*, 1264–1275. [CrossRef]
33. Konat, Ł.; Zemlik, M.; Jasiński, R.; Grygier, D. Austenite grain growth analysis in a welded joint of high-strength martensitic abrasion-resistant steel hardox 450. *Materials* **2021**, *14*, 2850. [CrossRef] [PubMed]
34. Konat, Ł. Technological, microstructural and strength aspects of welding and post-weld heat treatment of martensitic, wear-resistant Hardox 600 steel. *Materials* **2021**, *14*, 4541. [CrossRef]
35. Di, H.; Sun, Q.; Wang, X.; Li, J. Microstructure and properties in dissimilar/similar weld joints between DP780 and DP980 steels processed by fiber laser welding. *J. Mater. Sci. Technol.* **2017**, *33*, 1561–1571. [CrossRef]
36. Łętkowska, B.; Dudziński, W.; Frydman, S. Abrasive wear for selected grades of low-carbon boron steels at different states of heat treatment. *Q. Tribol.* **2012**, *4*, 151–166.
37. Liu, Y.; Liskiewicz, T.W.; Beake, B.D. Dynamic changes of mechanical properties induced by friction in the Archard wear model. *Wear* **2019**, *428–429*, 366–375. [CrossRef]
38. Hsu, S.M.; Shen, M.C.; Ruff, A.W. Wear prediction for metals. *Tribol. Int.* **1997**, *30*, 377–383. [CrossRef]
39. Napiórkowski, J.; Lemecha, M.; Konat, Ł. Forecasting the Wear of Operating Parts in an Abrasive Soil Mass Using the Holm-Archard Model. *Materials* **2019**, *12*, 2180. [CrossRef] [PubMed]
40. Li, C.; Deng, X.; Huang, L.; Jia, Y.; Wang, Z. Effect of temperature on microstructure, properties and sliding wear behavior of low alloy wear-resistant martensitic steel. *Wear* **2020**, *442*, 203125. [CrossRef]
41. Hokkirigawa, K.; Kato, K. An experimental and theoretical investigation of ploughing, cutting and wedge formation during abrasive wear. *Tribol. Int.* **1988**, *21*, 51–57. [CrossRef]
42. Rendón, J.; Olsson, M. Abrasive wear resistance of some commercial abrasion resistant steels evaluated by laboratory test methods. *Wear* **2009**, *267*, 2055–2061. [CrossRef]

materials

MDPI

Article

Processing of Haynes® 282® Alloy by Direct Energy Deposition with Arc and Wire

Manuela Zinke [1,*], Stefan Burger [2] and Sven Jüttner [1]

1 Institute of Materials and Joining Technology, Otto von Guericke University, 39106 Magdeburg, Germany
2 Bionic Production GmbH, 21339 Lueneburg, Germany
* Correspondence: manuela.zinke@ovgu.de

Abstract: Direct energy deposition with arc and wire (DED-AW) is a versatile, low-cost, and energy-efficient technology for additive manufacturing of medium- and large-sized metallic components. In this study, the effects of arc energy and shielding gas in cold metal transfer (CMT) welding of walls and blocks on cooling time, mechanical properties, and macro- and microstructure have been studied using precipitation-hardenable Ni-based superalloy Haynes® 282®. The arc energy and consequently the cooling rate were varied by changing the wire feed rate and the travel speed. As expected, increasing the arc energy leads to higher cooling times for the walls. Due to the 2D thermal conduction, the thin walls cool down much slower than multi-layer welded blocks, but this reduces the strength values only very slightly. While the walls have no sensitivity to the occurrence of unacceptable seam irregularities, the multi-layer blocks show isolated seam defects, such as hot cracks or lack of fusion. Despite shielding gas variation, the as-welded blocks show acceptable strength properties at room temperatures (RT) and impact values at RT and −196 °C. However, the use of an N-containing shielding gas results in lower elongation and notched bar impact energy. Precipitation-hardened specimens tested at 871 °C exhibit a similar strength level to transverse tensile specimens of gas metal arc welding (GMAW) welded joints on 12.7 mm thick plates with fracture in the weld metal.

Keywords: additive manufacturing; Haynes® 282®; CMT process; shielding gas; cooling time; mechanical properties; microstructure; PWHT

check for updates

Citation: Zinke, M.; Burger, S.; Jüttner, S. Processing of Haynes® 282® Alloy by Direct Energy Deposition with Arc and Wire. *Materials* **2023**, *16*, 1715. https://doi.org/10.3390/ma16041715

Academic Editors: Swarup Bag and Christ Prakash Paul

Received: 13 January 2023
Revised: 8 February 2023
Accepted: 16 February 2023
Published: 18 February 2023

1. Introduction

In additive manufacturing (AM), parts are made from 3D model data by depositing layer by layer of material [1]. It consists of seven process categories, including directed energy deposition (DED) and powder bed fusion (PBF), both of which enable the generation of near-full-density metal parts. In contrast to powder bed processes, in directed energy deposition (DED), the feedstock (powder or wire) is melted by means of an arc, laser beam, or electron beam and introduced directly into the molten pool [2]. In the literature, the DED arc and wire technology (DED-AW) is also often referred to as wire arc additive manufacturing (WAAM).

Modern high-temperature applications such as in gas turbines for the aerospace and energy sectors would not operate safely without utilizing Ni-based superalloys, which exhibit outstanding creep and oxidation resistance at elevated temperatures [3]. These superalloys are strengthened by solid-solution and precipitation hardening brought by two main types of precipitates: gamma-prime (γ') and gamma-double-prime (γ'') [4].

HAYNES® 282® (HY282, Ni-19Cr-10Co-8.5Mo-1.5Al-2.1Ti) is a precipitation-hardenable nickel-based superalloy developed for the above high-energy applications in the operating temperature range of 649 to 927 °C. It achieves its properties through a relatively low γ'-phase volume fraction of 19% and by solid-solution strengthening through additions of Mo, Cr and Co. In addition, Co also influences the solubility of the elements in the gamma matrix, Cr provides a high oxidation and hot corrosion resistance, and Mo is responsible for the excellent creep resistance. The microstructure of this alloy also includes MC, M_6C, and $M_{23}C_6$ carbides, which also play a role in strengthening [5,6]. The welds of HY282 also show excellent solidification and HAZ liquid cracking resistance, which has been demonstrated by Varestraint testing and Gleeble hot ductility tests [7]. Many studies [8–12] report on the successful processability of this material by conventional welding.

In terms of the additive manufacturing of HY282, the investigations focus on electron or laser beam powder bed fusion (PBF) [13–17] and DED with laser [18]. Few publications are available for WAAM with this alloy, although it is of interest for medium and large-size components in high-temperature applications of aircraft engines and industrial gas turbines. Thus, in [19,20], the influence of different build-up strategies for the gas tungsten arc welding (GTAW) of block and cylinder parts of this material on the recrystallization behavior and grain structures, as well as different heat treatment strategies on the carbide development and γ'-precipitation, were investigated. Investigations on the influence of different arc energies and shielding gas mixtures during additive CMT arc welding with HY282 on the mechanical properties and microstructure, on the other hand, are not yet available.

Cold metal transfer (CMT) is a modified GMAW process that provides low arc energy and controlled material deposition. It has a low dilution rate, less spattering, and high energy density and shows good suitability for additive manufacturing of wall structures with different Ni-based wire electrodes [21–24]. However, when producing CMT multi-track depositions with stringer beads, lack of fusion and hot cracks can occur [25]. To improve the wetting behavior of the viscous nickel melt and reduce the risk of lack of fusion, other GMAW modes of metal transfer, such as pulsed arc or mixed shielding gases with argon, helium, hydrogen, or carbon dioxide, can be used. He and H_2 ensure a higher arc temperature, which means that larger weld penetrations and higher travel speeds can be achieved. CO_2 can also be added in very small quantities (ppm range) to increase arc stability [26]. In addition, the waviness of side surfaces and the mechanical properties of walls can be influenced by the shielding gas [27].

Nitrogen is also used as a shielding gas component in the fusion welding of Ni-based superalloys to affect the weld metal metallurgically. In the case of the high-temperature superalloy Alloy 602 CA (NiCr25FeAlY), which has good high-temperature strength up to 1200 °C due to the precipitation of primary Cr carbides, the addition of nitrogen in the shielding gas leads to a significant reduction in the hot cracking tendency in weld metal [28,29]. Nitrogen exerts a positive effect on carbide morphology in unidirectional and single-crystal Ni-based alloys, as reported by [30,31]. Furthermore, with increasing N content in the shielding gas, an increase in room temperature strength and hardness of the GTAW weld metal of Alloy 263 and Alloy X can be observed [32].

Therefore, the present paper informs about the influence of arc energy and shielding gas during WAAM of wall and block parts with HY282 on cooling times, macro- and microstructures as well as mechanical-technological properties at room temperature and elevated temperature. The tests were performed in the welded and partially precipitation-hardened condition. Different geometries were used to generate various welding conditions (one bead/layer walls or multi-layer deposits with stringer beads) and residual stress conditions. The weld deposits were carried out with different wire diameters of HY282, since not enough material was available in one diameter.

2. Experimental Setup

2.1. Filler Metal

A batch of the solid wire electrode HY282 in two different diameters (0.040″ and 0.035″) was available for the investigations. Table 1 shows the target chemical composition by the wire manufacturer [33] and the actual chemical composition of the wire batch used.

Table 1. Chemical composition of solid wire electrodes HY282 (wt%).

Analysis	Φ, ″	C	Si	Mn	S	Cr	Fe	Mo	Co	Al	Ti	B	Ni
Target	-	0.04–0.08	≤0.15	≤0.3	≤0.015	18.5–20.5	≤1.5	8–9	9–11	1.38–1.65	1.9–2.3	0.003–0.010	bal.
Actual	0.040	0.065	0.05	0.05	0.0006	19.23	0.99	8.15	10.05	1.59	2.20	0.0039	57.45
	0.035	0.065	0.05	0.05	0.0006	19.23	0.99	8.16	10.10	1.59	2.26	0.0040	57.32

This was determined using atomic emission spectrometry with Spectrolab S (SPECTRO Analytical Instruments GmbH) on buttons melted from the solid wire electrodes using the Arc Melter MAM-1 (Edmund Bühler GmbH) under 100% argon. The methodology is described in [34]. The contents of C and S were determined on the as-received wire electrodes by combustion analysis using the G4 ICARUS Series 2 (Bruker Corporation).

2.2. Setup for WAAM

In this study, the cold metal transfer (CMT) process was selected for gas metal arc welding (GMAW) because of its low arc energy and stable process behavior. The deposits were made in synergic mode using power source CMT Advanced 4000 with RCU 5000 control unit and a three-axis gantry machine with programmable logic controller. All weld beads were produced in flat position (1G) with neutral welding torch orientation and alternating welding direction after each layer. No brushing was done between the layers and the maximal interpass temperature was 100 °C. The electrical welding parameters, the shielding gas flow, and the wire feed speed were additionally recorded with the external measuring system Weld Analyst-S2 (HKS Prozesstechnik GmbH). The cooling time measurements were made with a 1-channel pyrometer (measuring range: 385 to 1600 °C). They were measured directly behind the arc on the seam surface of every second layer of the walls and every second weld bead of the blocks. Due to the absence of a specific temperature interval for Ni-based materials (comparable to the $t_{8/5}$ cooling time for fine-grain structural steels or the $t_{12/8}$ cooling time for CrNi steels), the times for cooling from 1000 to 600 °C were determined. In this range, secondary phases precipitate in many Ni-based alloys. Table 2 gives an overview of the experimental program.

Table 2. Experimental program for generation of different WAAM parts with HY282 with variable arc energy and shielding gases.

Step	Criterion	Φ (″)	CMT Reference No.	Shielding Gas (acc. to [35])	Setting Values (m/min) WFS	TS	Structure	Dimensions (mm) L	H	W
1.	Arc energy	0.040	1254 V2.3.8.4	Z-ArHeHC 30/2/0.05	8.0; 11.0; 14.0	0.4; 0.6; 0.8	wall	225	65	variable
	Structure	0.040	1254 V2.3.8.4	Z-ArHeHC 30/2/0.05	11.0	0.6	block	330	32	55
2.	Shielding gas	0.035	960 V1.0.0.3.4	Z-ArHeHC 30/2/0.05 R1-ArHeH30/1	12.5	0.4	wall block	180 220	60 27	variable 30
			1699 V4.5.0	Z-ArHeNC 5/5/0.05	12.5	0.55	wall block	180 220	60 28	variable 35

In the first step, the influence of arc energy on the cooling rate, the macro- and microstructure, and the mechanical properties at room temperature (RT) during welding of thin walls was investigated. For this, the wire feed speed (WFS) was varied between 8.0 and 14.0 m/min and the travel speed (TS) between 0.4 and 0.8 m/min. The constant wall height resulted in a different number of beads and wall widths. A selected arc energy was used to produce a multi-layer block with the stringer bead technique.

In the second step, the influence of shielding gases on the abovementioned properties was investigated. Additional to the shielding gas Z-ArHeHC 30/2/0.05, which contains Ar, 30% He, 2%H_2, and 550 ppm CO_2, the reducing shielding gas R1-ArHeH30/1 and the

5% N_2-containing shielding gas Z-ArHeNC 5/5/0.05 were used for the reasons mentioned above.

3. Test Methods

The various components were first subjected to a visual and dye penetration test. Metallographic cross sections (Y–Z plane) were then taken. For the automatic preparation, it was necessary to cut the cross sections of walls and blocks in the middle. All samples were processed with standard metallographic techniques with a final polish of 0.5 µm O.P.S (SiO_2). Subsequently, the central part of the walls and blocks were mechanically machined to allow high-quality X-ray inspection. While the image quality indicator according to [36] on the walls was 16 or 17, only an indicator of 10 was achieved on the blocks due to their higher thickness. For this reason, no radiographic testing was performed on the blocks. On the walls and blocks, chemical analyses, tensile tests, and hardness measurements were performed. Standard tensile tests were performed to BS EN ISO 6892-1 [37] at room temperature. The tensile specimens with form E 5 × 10 × 40 [38] were taken only in welding direction. Vickers hardness testing was measured under a test load of 10 kg and a load dwell time of 8 s in compliance with BS EN ISO 6507-1 [39]. Only on the blocks was the determination of the impact energy on Charpy V specimen at room and low temperatures (−196 °C) carried out. The microstructure of different WAAM parts was investigated using a light optical microscope (Inverse incident light microscope Leica MeF4A, Leica) and scanning electron microscope (XL30 FEG/ESEM, company FEI/Philips).

The post-weld heat treatment (PWHT) was performed in a vacuum chamber furnace from IVA Schmetz GmbH in conjunction with a thermo process management system in a vacuum on final machined specimens. It followed the standard procedure and was performed at 1160 °C/2 h + 1010 °C/2 h + 788 °C/8 h. For accelerated cooling, argon was injected into the furnace and swirled by a large speed-controlled fan. The hot tensile tests at 871 °C were performed with the test machine Zwick Roell Z100 with a 3-zone furnace at atmosphere in compliance with the ISO 6892 [40].

4. Results

4.1. Process Parameters and Arc Energy

Table 3 shows measured arithmetical values for wire feed rate (WFS), current (I), and voltage (U) as a function of the setting values when deposited walls and block with the standard shielding gas Z-ArHeHC 30/2/0.05. Arc energy (En) is determined in welding as the ratio of arc power (P) to arc travel speed (TS), as in Equation (1).

$$\mathrm{En} = \frac{\overline{\mathrm{P}}}{\mathrm{TS}} = \frac{\overline{\mathrm{I}} \cdot \overline{\mathrm{U}}}{\mathrm{TS}} \tag{1}$$

Table 3. Process parameters of different WAAM parts with HY282 (Φ 0.04″) and shielding gas Z-ArHeHC 30/2/0.05 (flow rate: 18 L/min; stick-out: 12 mm).

Process Parameter			Wall									Block
Setting	WFS	m/min	8.0	8.0	8.0	11.0	11.0	11.0	14.0	14.0	14.0	11.0
	TS	m/min	0.4	0.6	0.8	0.4	0.6	0.8	0.4	0.6	0.8	0.6
Measured	$\overline{\mathrm{WFS}}$	m/min	6.5 ± 0.2	7.1 ± 0.5	7.4 ± 0.3	9.3 ± 0.3	10.5 ± 0.4	11.0 ± 0.4	13.9 ± 0.4	14.2 ± 0.2	14.1 ± 0.2	10.7 ± 0.5
	$\overline{\mathrm{I}}$	A	117 ± 3.4	124 ± 4.2	127 ± 3.1	145 ± 3.5	153 ± 6.5	152 ± 8.6	181 ± 5.6	188 ± 7.4	188 ± 7.8	159 ± 8.2
	$\overline{\mathrm{U}}$	V	17.3 ± 0.9	19.6 ± 2.0	19.8 ± 2.2	20.9 ± 0.8	21.2 ± 1.6	22.0 ± 2.5	23.1 ± 0.7	23.8 ± 1.0	24.0 ± 1.2	20.6 ± 2.0
	$\overline{\mathrm{En}}$	J/mm	303 ± 12	242 ± 28	189 ± 21	455 ± 17	323 ± 20	249 ± 24	625 ± 25	447 ± 16	338 ± 13	327 ± 27

The arc energy (En) given in Table 3 was calculated as the arithmetic mean value of the arc energies of the single weld beads. Differences between the nominal and actual wire feed rates occurred for all welds, which affects the deposition rate. This results from the fact that the CMT synergic lines used were not developed for the used wire electrode–shielding gas combinations. In order to achieve a stable process behavior, the welding power source uses an internal control to regulate, for example, the wire feed frequency, which affects the average of the wire feed speed. The deposition rate (DR) is the amount of wire melted per time. At constant wall height, the wall width, the number of layers, and the deposition rate increase with increasing heat energy (Figure 1).

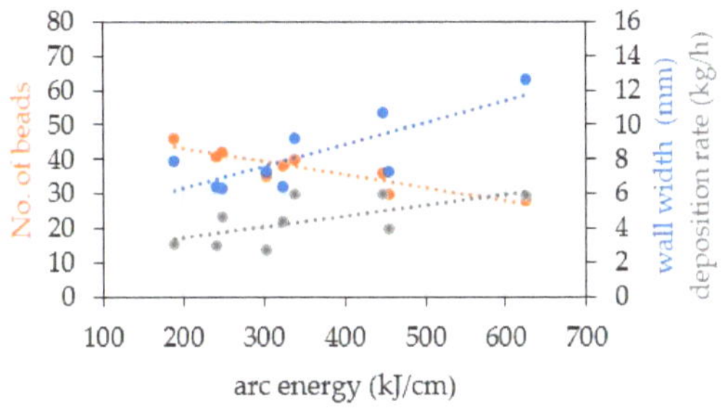

Figure 1. Influence of arc energy on number of layers, walls width, and deposition rate when welded the walls with variable process parameters from Table 2.

The shielding gas tests, implemented with the thinner 0.035″ wire diameter of HY 282, required the choice of another CMT reference No. 960 V1.0.0.3.4. However, the use of the nitrogen-containing shielding gas with only 5% helium reduced the arc voltage to such an extent that a more unsteady arc and an unacceptable weld geometry occurred. For this reason, the CMT reference No. 1699 V.4.5.0 was selected for this shielding gas. In order to ensure approximately the same arc energy and weld geometries, the welding speed was increased from 0.4 to 0.55 m/min. As can be seen in Figure 2, the cross sections of single beads show a similar seam geometry. Only the penetration depth increases significantly when using CMT 1699.

Shielding gas	Z-ArHeHC 30/2/0.05	Z-ArHeNC 5/5/0.05	Z-ArHeNC 5/5/0.05
Synergic characteristic	CMT 960	CMT 960	CMT 1699
WTS / TS (m/min)	12.5 / 0.4	12.5 / 0.4	12.5 / 0.55
Width w / heigth h (mm)	6.8 / 2.8	6.8 / 3.5	6.9 / 2.7
Penetration depth d_p (mm)	0.35	0.57	4.3
Wetting angle θ (°)	84.4	85.9	85.4

Figure 2. Seam geometries of single beads on the plate (Y–Z plan) welded with different shielding gases and CMT lines.

Table 4 shows the arithmetic mean values for process parameters when welding with different CMT synergic lines and shielding gases. The higher average wire feed speeds and welding currents result from the CMT 1699 synergic line, which differs significantly in its electrical waveform from the CMT 960 synergic line.

Table 4. Process parameters of different WAAM structures with HY282 (Φ 0.035″) and variable CMT synergic lines and shielding gases (flow rate: 18 L/min; stick-out: 12 mm).

Shielding Gas					Synergic Line Short Name	Structure	Setting Values		Measured Values				
Ar	He	H_2	N_2	CO_2			WFS	TS	$\overline{WFS}$	$\overline{I}$	$\overline{U}$	$\overline{En}$	DR
Bal.	30	2	-	0.055	CMT 960 ArHeHC	wall	12.5	0.40	10.0 ± 0.3	114 ± 1.8	18.0 ± 0.4	309 ± 11	3.1
						block	12.5	0.40	11.0 ± 1.0	118 ± 1.7	18.2 ± 1.4	321 ± 27	3.4
Bal.	30	1	-	-	CMT 960 ArHeH	wall	12.5	0.40	11.7 ± 0.7	115 ± 1.1	19.3 ± 0.7	333 ± 12	3.6
						block	12.5	0.40	11.1 ± 1.1	118 ± 1.7	18.6 ± 1.3	322 ± 23	3.4
Bal.	5	-	5	0.055	CMT 1699 ArHeNC	wall	12.5	0.55	14.1 ± 0.3	151 ± 3.2	21.5 ± 04	356 ± 8.9	4.4
						block	12.5	0.55	13.5 ± 0.9	152 ± 2.6	20.4 ± 1.1	339 ± 14	4.1

A CMT cycle consists of a plasma phase (composed of boost and wait stage) and the short-circuit phase. In the boost stage, the filler wire and base material are melted, producing the molten droplet and the weld pool. During the short-circuit phase, the electromagnetic pinch effect causes the molten droplet to constrict at the tip of the wire electrode. For the CMT 960 synergic line, this process is supported by a small current pulse of about 120 A (Figure 3a). In contrast, with the CMT 1699 synergic line, the welding current initially drops briefly in the boost state, then rises and ends with a targeted, almost vertical peak of approx. 380 A (Figure 3b). In addition to that, the short-circuit time is significantly shorter here, and the frequency oscillating wire feeding (f_{WFS}) thus the number of CMT cycles per second is higher. This causes the higher welding currents and arc voltages.

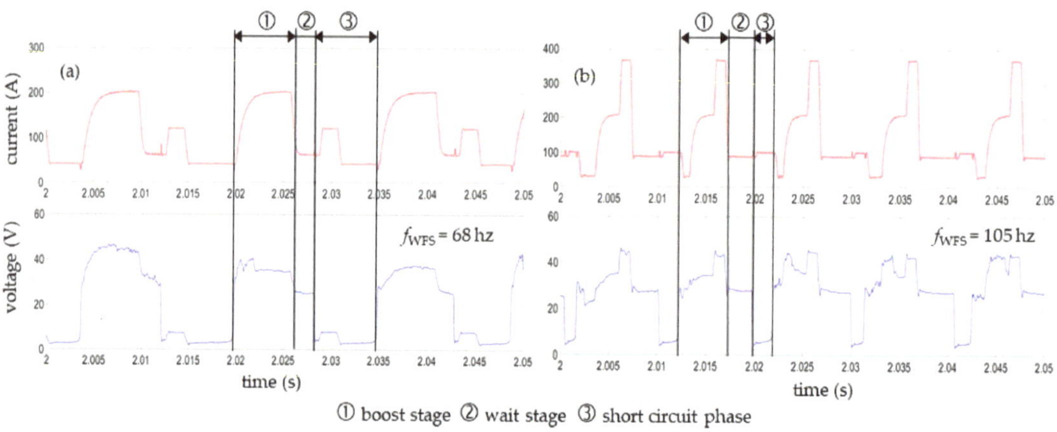

Figure 3. Typical electrical waveform of the CMT cycles for (**a**) CMT 960 and (**b**) CMT 1699.

4.2. Cooling Times Depended on Process Parameters

Figure 4 shows the cooling times $t_{10/6}$ for the walls, depending on arc energy and layer. In the first layers, shorter cooling times occur due to the still-possible heat conduction into the substrate. Starting from the ninth layer, the 3D heat conduction changes to a 2D heat conduction, which is why the cooling time remains almost constant. As expected, the cooling times increase with increasing arc energy when welding the walls. At the highest arc energy of 625 J/mm, the cooling time is about 24 s. At the lowest arc energy of 189 J/mm, the cooling time is only 8 s. In the case of the multi-pass block, the average cooling time is only approx. 4.1 s due to the 3D heat conduction. It is thus significantly shorter than that of the bar with comparable arc energy.

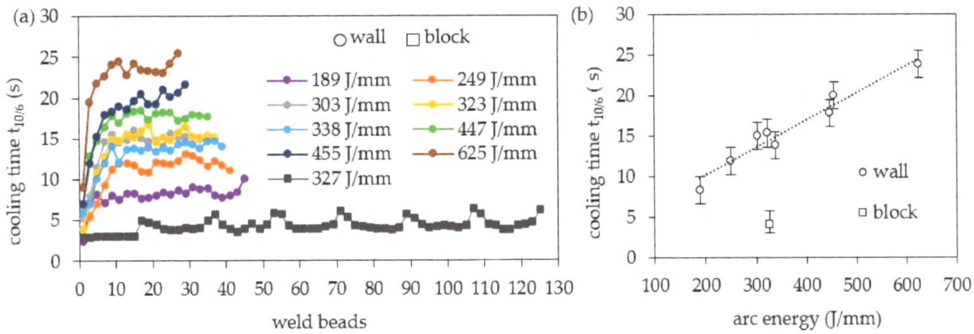

Figure 4. Cooling times of different parts (wall, block) welded with HY282 (Φ 0.04″) and Z-ArHeHC 30/2/0.05 (**a**) depending on arc energy and weld beads and (**b**) arithmetic mean values at 11th weld bead depending on arc energy.

Independent of the structure, the shielding gas shows only a minor influence on the cooling time since the values for arc energy do not differ significantly (Figure 5).

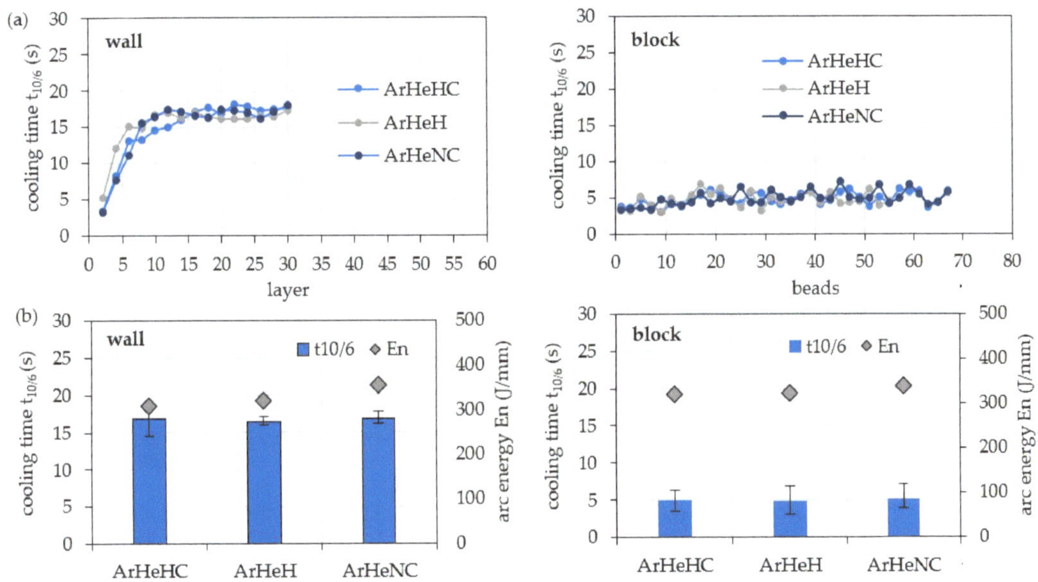

Figure 5. Cooling times of different parts (wall, block) welded with HY282 (Φ 0.035″) depending on shielding gas. (**a**) Single values over the layers. (**b**) Arithmetic mean values from 10th layer.

4.3. Non-Destructive Testing and Chemical Analysis

Visual and penetrant inspections of all printed walls and blocks did not reveal any external defects such as cracks, lack of fusion, or gas pores in the weld metal. Figure 6 shows two walls with low and high arc energy and the 330 mm long block. The wall produced with the higher arc energy has a dull, slightly more oxidized surface than the wall produced with the lower arc energy because the melt pool dimensions and cooling times are higher here. Figure 7 shows the walls and the blocks produced with the different shielding gases. The walls, which are still unbrushed, show the metal vapor deposits with different colors resulting from the welding process. The block produced with the N shielding gas shows inhomogeneous fusion lines between the single stringer beads.

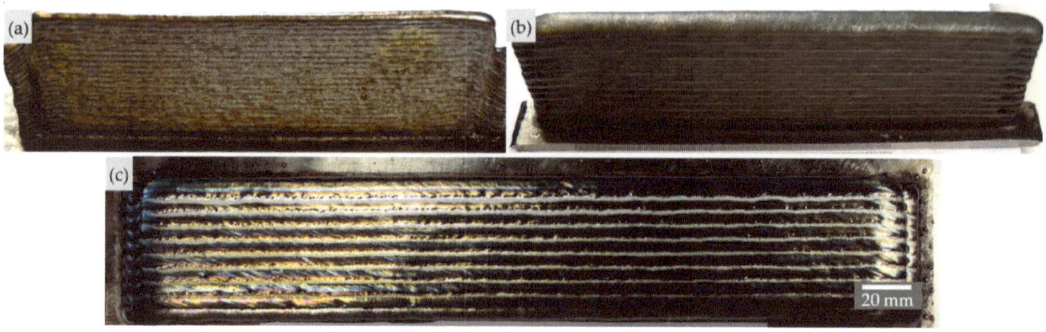

Figure 6. Side views of brushed walls. (**a**) WFS: 8.0 m/min, TS: 0.8 m/min; (**b**) WFS: 11.0 m/min, TS: 0.6 m/min; (**c**) top view of the unbrushed block with WFS: 14.0 m/min, TS: 0.4 m/min (Z-ArHeHC).

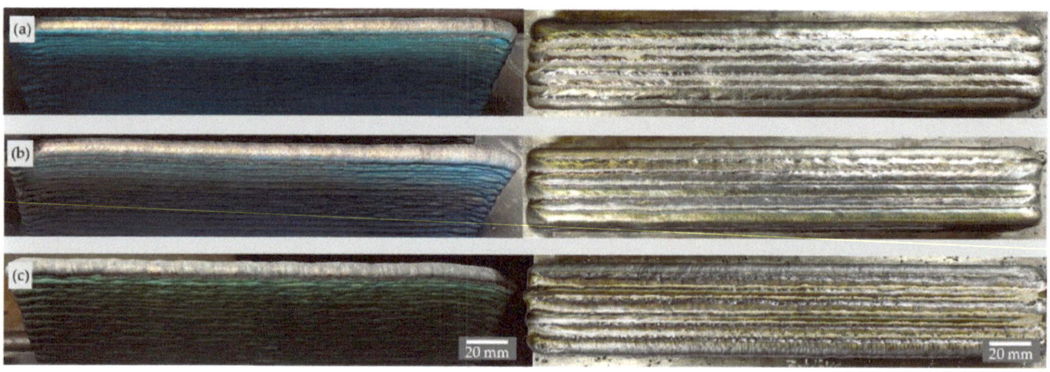

Figure 7. Side view of the unbrushed walls (**left**) and top views of the brushed blocks (**right**). (**a**) Z-ArHeHC; (**b**) R1-ArHeH; (**c**) Z-ArHeNC (welding parameters in Table 4).

The X-ray examinations of all walls showed no or only very small gas porosity of less than 0.001%. No radiographic tests were performed on the blocks due to the low image quality number as they did not provide sufficient information.

Since a shielding gas with low active and reducing shielding gas components was used and no brushing between the individual layers took place, the chemical compositions of weld metals were determined. For the evaluation, arithmetic mean values were used for the individual alloying elements of the various wall and block structures, which were produced with varying arc energy and shielding gases. As shown Figure 8, no pickup or burning loss of alloying elements has occurred.

Carrier gas melt extraction (G8 Galileo, Bruker Corporation) was also used to measure the contents of oxygen and nitrogen of walls and blocks. The difference in gas content between weld metal and the wire electrode was determined using Equation (2).

$$\Delta_X = X_{weld\ metal} - X_{wire} \tag{2}$$

X_{wm}: Content of O or N in weld metal (wm)

X_{wire}: Content of O or N in wire electrode (wire)

In general, a slight increase in oxygen and nitrogen can be seen for all weld metal types. However, there is no significant influence of the arc energy (Figure 9a). On the other hand, the use of the shielding gas Z-ArHeNC 5/5/0.05 causes a significant increase in the nitrogen content of the weld metal (Figure 9b). The oxygen values are also slightly higher (Figure 9c).

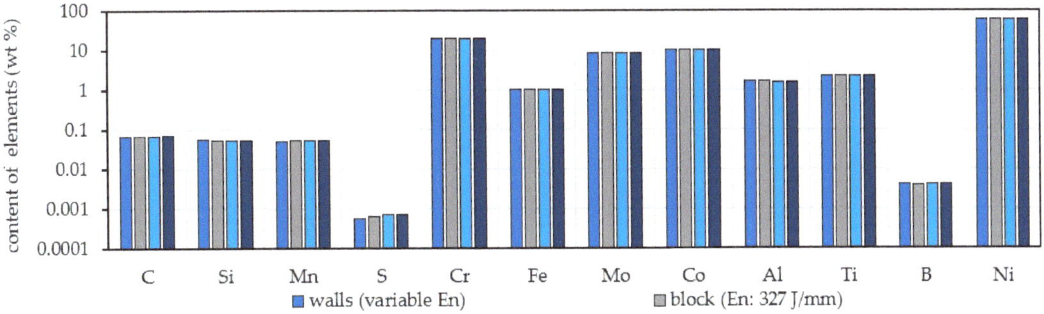

Figure 8. Arithmetic mean values of chemical compositions of HY282 walls and blocks depending on arc energy and shielding gas (scale of y-axis: log to base 2).

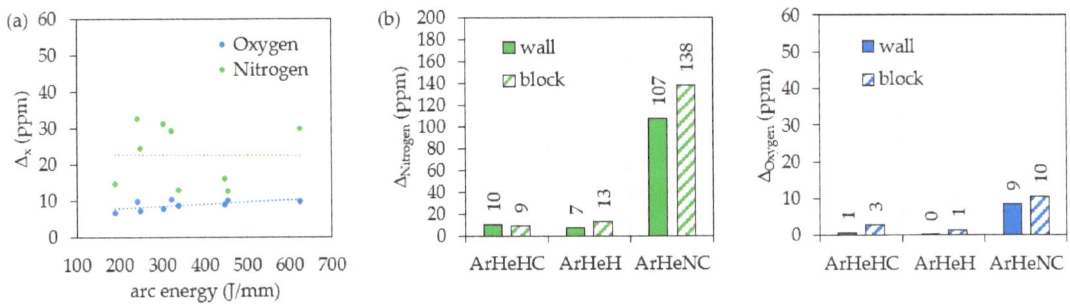

Figure 9. Increase of oxygen and nitrogen in the walls and blocks compared to the wire electrode HY282 depending on (**a**) arc energy (Φ 0.040″) and (**b**) shielding gas (Φ 0.035″).

4.4. Macrostructure

The cross sections of the walls in Figure 10 showed no unacceptable internal seam irregularities such as lack of fusion, pores, or hot cracks. In the multi-pass block, however, three micro hot cracks occurred with a total crack length of 810 μm and an *MSI* of 0.47 μm/mm^2.

The *MSI* (micro fissure sensitivity index) is the quotient of total micro fissure length and the weld metal area. Comparable block welds with Ni alloy Inconel 718 showed a significantly higher hot cracking occurrence with an *MSI* between 2.32 and 33.3 [25].

The macro sections of the walls and blocks deposited with different shielding gases showed similar behavior (Figure 11) The walls exhibited no seam defects, but in the blocks, unacceptable lack of fusion defects and a few hot cracks occurred. The lack of fusion was caused by the bulging stringer weld beads, the inhomogeneous fusion lines, and the excessive lateral welding torch feed. The shielding gas influenced the hot cracking occurrence in the blocks as follows:

- Z-ArHeHC: six hot cracks, 1080 μm total crack length, 2.50 μm/mm^2 MSI
- R1-ArHeH: two hot cracks, 390 μm total crack length, 1.05 μm/mm^2 MSI
- Z-ArHeNC: no hot cracks.

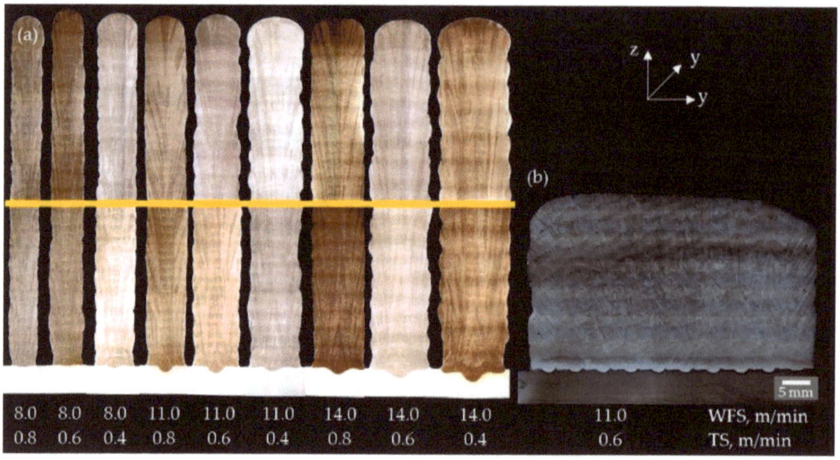

Figure 10. Macro cross sections (Y–Z plane) of walls (**a**) and block (**b**) welded with HY282 (Φ 0.04″) depending on arc energy (etched with Beraha III).

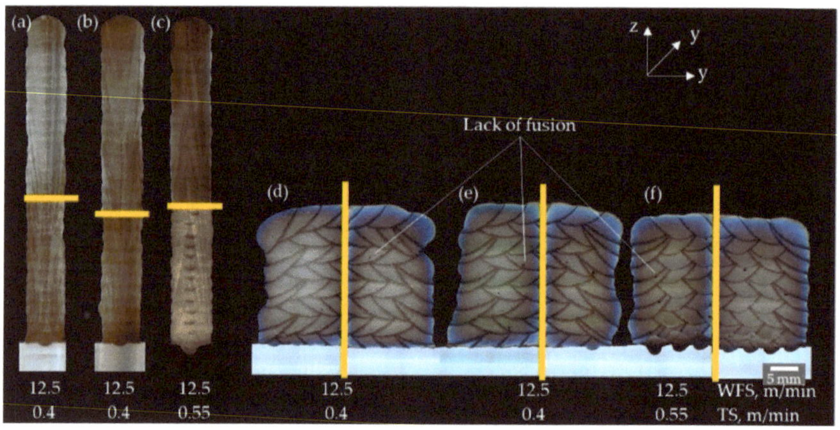

Figure 11. Macro sections (Y–Z plane) of walls (**left**) and blocks (**right**) with HY282 (Φ 0.035″) depending on shielding gases. (**a**,**d**) ArHeHC; (**b**,**e**) ArHeH; (**c**,**f**) ArHeNC (wall: etched with Beraha III, block: electrolytic etched with 10% CrO_3 and 90% water).

In order to determine the influence of the welding parameters on the surface waviness of the walls, the machining allowance *MA* on the cross sections (Y–Z plane) of the walls was determined by image analysis. This allowance would be required to mill a flat structure from the wall. It is therefore the dimensional difference between the finished dimension and the initial dimension of the wall width, e.g., for subsequent milling. It can be seen that the surface waviness increases with increasing wire feed speeds. Figure 12 shows that higher arc energy has an unfavorable effect on the surface waviness. With respect to the shielding gases influence, the walls manufactured with ArHeNC and CMT1699 synergic line have the lowest waviness.

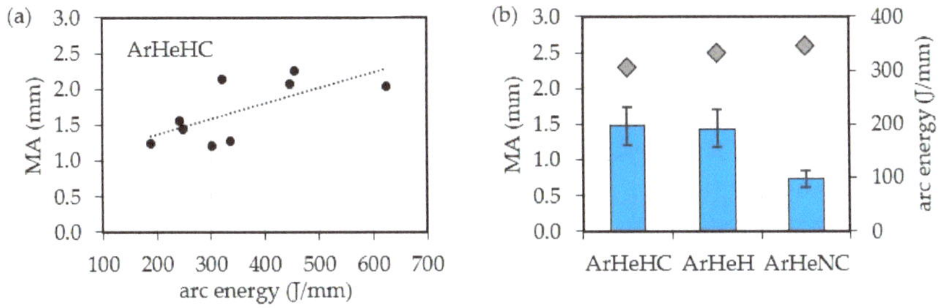

Figure 12. Machining allowance *MA* of walls depended on (**a**) arc energy and (**b**) shielding gas.

Figure 13 exemplifies the microstructures of the last deposited layer of the walls produced with different shielding gases. The additive weld metal exhibits the typical dendritic solidification microstructure with interdendritic precipitated MC described in the literature [8,9] for welded HY282. Additionally, isolated large orange particles are visible which are classified as Ti(C,N) on the basis of their color and geometry. There are many small angular yellow particles which are probably MN-type nitrides or MCN-type carbonitrides, where M can stand for Ti, Al, and/or B due to the given chemical composition of HY282. These elements are known to have a high affinity for nitrogen and optionally carbon. As expected, the amount of these particles is greater in the weld metal deposited with ArHeNC due to the higher N content.

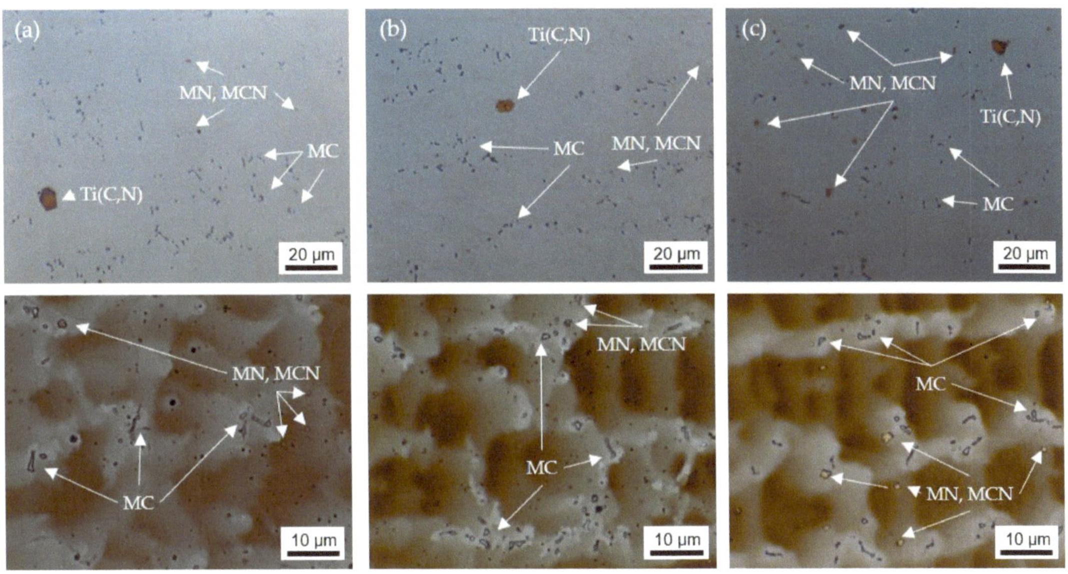

Figure 13. Micro sections (Y–Z plane) of walls with HY282 (Φ 0.035″). (**a**) ArHeHC; (**b**) ArHeH; (**c**) ArHeNC (**top**: polished, **bottom**: etched with Beraha III).

4.5. Mechanical Properties at Room Temperature

Tables 5 and 6 show the results of quantitative tensile tests at room temperature (RT) for the wall and block parts depending on the arc energy and the shielding gas used.

Table 5. Mechanical properties of wall and block parts of HY282 (Φ 0.04″) depending on arc energy.

Symbol	Unit	Wall									Block
WFS TS	m/min m/min	8.0 0.4	8.0 0.6	8.0 0.8	11.0 0.4	11.0 0.6	11.0 0.8	14.0 0.4	14.0 0.6	14.0 0.8	11.0 0.6
UTS	MPa	832 ± 4	807 ± 26	833 ± 5	810 ± 3	851 ± 49	802 ± 12	849 ± 26	840 ± 12	821 ± 9	879 ± 10
0.2% YS	MPa	540 ± 3	514 ± 15	528 ± 2	523 ± 22	563 ± 24	525 ± 8	549 ± 8	567 ± 14	558 ± 5	594 ± 29
A	%	40.1 ± 2.5	42.3 ± 1.1	44.2 ± 0.5	41.6 ± 1.6	43.8 ± 0.7	43.6 ± 1.1	41.8 ± 1.2	39.7 ± 2.4	37.7 ± 5.1	43.8 ± 1.7
Hardness	HV10	267 ± 9	264 ± 9	267 ± 9	261 ± 7	276 ± 11	268 ± 8	263 ± 11	267 ± 9	277 ± 13	271 ± 13

Table 6. Mechanical properties of wall and block parts of HY282 (Φ 0.035″) depending on shielding gas.

Structure	Parameter	Unit	ArHeHC	ArHeH	ArHeNC
wall	UTS	MPa	810 ± 30	796 ± 25	799 ± 16
	0.2% YS	MPa	516 ± 13	517 ± 22	527 ± 19
	A	%	46.0 ± 1.8	44.0 ± 5.0	31.1 ± 4.1
	Hardness	HV10	276 ± 16	265 ± 8	293 ± 13
block	KV_2 at RT	J	144 ± 4	142 ± 3	133 ± 9
	KV_2 at −196 °C	J	133 ± 7	129 ± 1	126 ± 0
	Hardness	HV10	278 ± 18	270 ± 14	296 ± 18

As can be seen in Figure 14, the strengths (UTS, YS) and the primary dendrite arm spacing (PDAS) increase slightly with increasing arc energy, i.e., with slower cooling. On the other hand, the elongation at break is slightly reduced. The block shows higher tensile and 0.2% yield strength at the same setting values (WFS: 11.0 m/min, TS: 0.6 m/min) despite faster cooling. Only this structure achieves the typical tensile data specified by the manufacturer for all weld metal (UTS: 860 MPa, 0.2% YS: 586 MPa, A: 40.0%) [41]. However, the scatter of the single values (see Table 5) should also be considered when making these observations. With regard to the shielding gases, it should be noted that the use of ArHeHC and ArHeH lead to comparable mechanical properties. The use of nitrogen in the shielding gas causes increased hardness, lower Charpy V values, and a significant reduction in elongation (see Table 6). Here is an influence of the precipitated MN-type nitrides or MCN-type carbonitrides suspected.

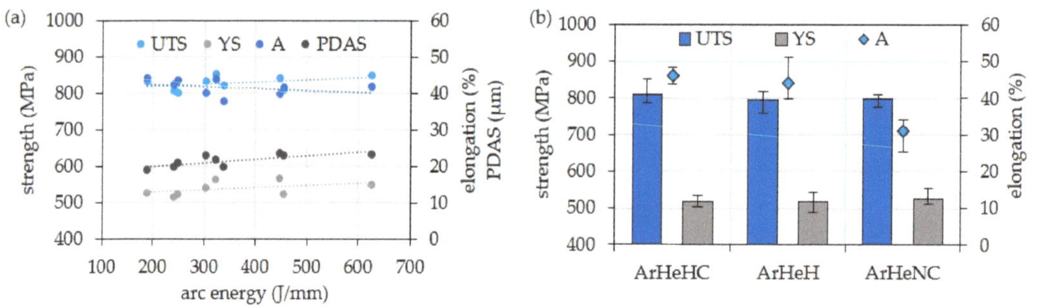

Figure 14. Results of tensile tests at RT of walls of HY282 depended on (**a**) arc energy and (**b**) shielding gas.

4.6. Precipitation Hardening

Post-weld heat treatment (PWHT) was performed on specimens from the blocks welded with variable shielding gases. Since the cross sections of these blocks show lack of fusion defects, 3 mm thick flat specimens were first eroded from the block parts, which were then X-rayed. Figure 15a shows hot tensile specimen dimensions, and Figure 15b shows an exemplary radiographed flat specimen. The tensile specimens were then arranged in a way that the test resides in a defect-free part of the specimen.

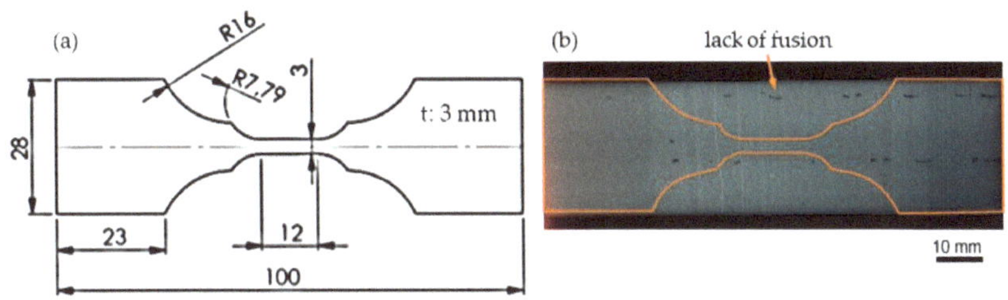

Figure 15. (**a**) Geometry of hot tensile specimens. (**b**) X-ray images of a flat specimen of block with Z-ArHeNC-5/5/0.05 digitized with a reflex camera.

The results of tensile testing at 871 °C of post-welding heat treated specimens are presented in Figure 16. Three specimens were tested in each case. The weld metal produced with Z-ArHeHC shows the highest strength. In contrast, the strengths of the weld metal produced with the nitrogen-containing shielding gas are slightly lower. In the latter case, however, the values are more scattered. All values reach the strength specified by the material manufacturer [41] (UTS: 565 MPa and 0.2% YS: 490 MPA) for a 12.7 mm plate welded with GMAW and a wire with a diameter of 0.045″ (fracture location: weld).

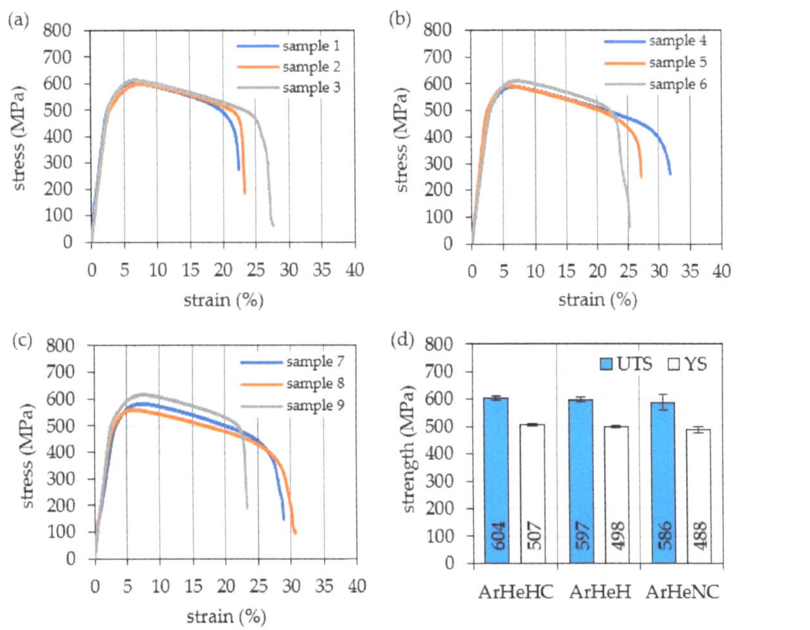

Figure 16. Results of hot tensile tests at 871 °C. (**a**) ArHeHC; (**b**) ArHeH; (**c**) ArHeNC; (**d**) comparison of the arithmetical mean values.

A metallographic evaluation was only possible for specimens produced with the standard gas since no material was available from the other samples. Figure 17 shows the microstructure in the as-built and precipitation hardened stage. While the carbides are primarily interdendritic in the post-deposited state, PWHT also causes precipitation of many discrete carbides at the grain boundaries and within the austenite grains. The appearance of the carbides is similar to the (Ti,Mo)C (as-built) and $(Cr, Mo)_{23}C_6$ (precipitation-hardened) type carbides described in [8,9,19,42].

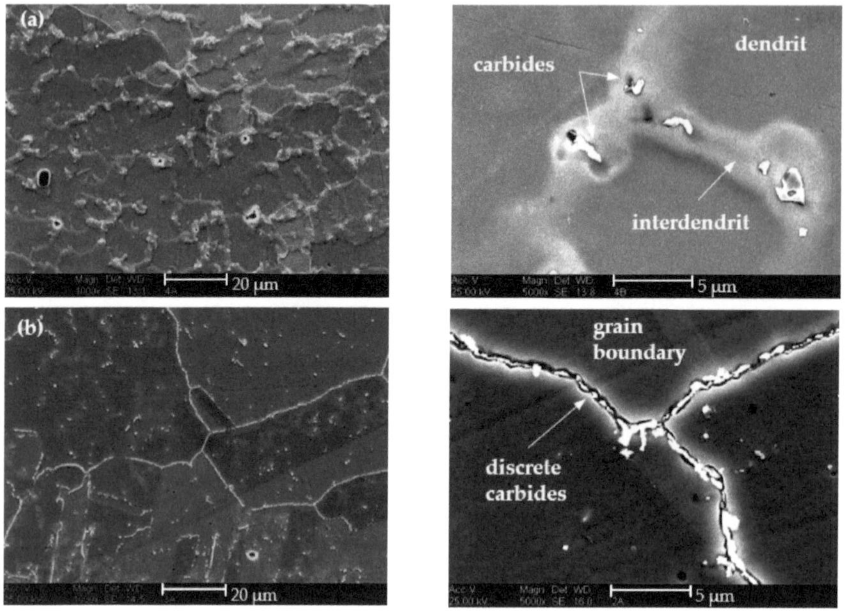

Figure 17. SEM micrographs of specimens, welded with shielding gas Z-ArHeHC. (**a**) as-built; (**b**) precipitation-hardened (etching solution: 5 g $C_2H_2O_4$ dissolved in 95 mL HCl).

In addition to the carbides, precipitation of the γ-phase occurs, as expected (Figure 18). A large number of these γ′-particles are in the size range between 60 nm to 80 nm. The precipitation-hardened weld metal has a higher hardness (346 HV10) than the weld metal (278 HV10).

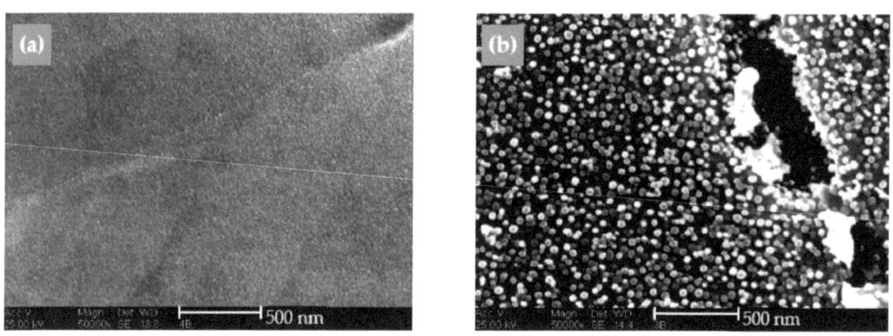

Figure 18. SEM micrographs of specimens, welded with shielding gas Z-ArHeHC. (**a**) as-built; (**b**) precipitation hardened (etching solution: 15 g CrO_3 + 150 mL H_3PO_4 + 100 mL H_2SO_4).

5. Conclusions

1. The results show the promising potential of nickel-based superalloy HAYNES® 282® for the WAAM of voluminous parts of medium to low complexity. The problem of the occurrence of hot cracks during arc welding, which is typical for fully austenitic Ni-alloys, only manifests itself to a limited extent in this alloy. It is only observed within the multi-track blocks due to the stronger thermo-mechanical reactions compared to the walls. However, compared to Inconel 718, it is significantly lower.

2. The CMT process showed a very good suitability for additive welding of walls (one bead/layer) of HY282. However, with the multi-track blocks there is a risk of the occurrence

of lack of fusion defects. Here, the use of the CMT impulse process is recommended because it leads to flatter weld beads with a lower wetting angle.

3. The mechanical properties at RT of the produced WAAM weld metal of the walls and blocks are slightly influenced by both the arc energy and the shielding gas. A higher arc energy results in slightly greater strength values and lower elongation. Within the shielding gases the N-containing shielding gas (Z-ArHeNC 5/5/0.05) results in higher hardness and in lower notched bar impact energy and elongation.

4. The results of tensile testing at 871 °C of post-welding heat treated specimens give good strength properties independent of the shielding gas. They all achieve the strength specified by the material manufacturer [41]. However, the weld metal produced with Z-ArHeHC shows the highest strength. In contrast, the strengths of N-containing weld metal are slightly lower.

5. The carbides observed in the weld metals correspond in appearance and location to those described in [8,9,19,42]—(Ti,Mo)C in the as-built condition and $(Cr, Mo)_{23}C_6$ in the precipitation hardening condition. In addition, large block-like nitrides and small nitride precipitates are visible. The latter occur significantly more frequently in the weld metal of Z- Z-ArHeNC 5/5/0.05.

Author Contributions: Investigation, S.B.; Writing—original draft, M.Z.; Funding acquisition, M.Z.; Writing—review and editing, S.J. All authors have read and agreed to the published version of the manuscript.

Funding: The present contribution is a part of the AiF project IGF-No. 20.616 BR/DVS-No. 01.3145 of the German Research Association on Welding and Allied Processes (DVS). It was kindly supported by the AiF (German Federation of Industrial Research Associations) within the program for promoting the Industrial Collective Research (IGF) of the Economic Affairs and Climate Action (BMWK), based on a decision by the German Bundestag.

Institutional Review Board Statement: Not applicable.

Informed Consent Statement: Not applicable.

Data Availability Statement: The data presented in this study are available in the article.

Conflicts of Interest: The authors declare no conflict of interest. The funders had no role in the design of the study; in the collection, analyses, or interpretation of data; in the writing of the manuscript; or in the decision to publish the results.

References

1. *DIN EN ISO/ASTM 52900:2022-03: Additive manufacturing-General Principles-Fundamentals and Vocabulary*; Beuth Verlag: GmbH, Berlin, 2022. [CrossRef]
2. Uriondo, A.; Esperon-Miguez, M.; Perinpanayagam, S. The present and future of additive manufacturing in the aerospace sector: A review of important aspects. *Proc. Proc. Inst. Mech. Eng. Part G J. Aerosp. Eng.* **2015**, *229*, 2132–2147. [CrossRef]
3. Mouritz, A.P. *Introduction to Aerospace Materials*; Mouritz, A.P., Ed.; Woodhead Publishing: Sawston, UK, 2012; pp. 1–14. [CrossRef]
4. Reed, R.C. *The Superalloys: Fundamentals and Applications*; Cambridge University Press: Cambridge, UK, 2006. [CrossRef]
5. Pike, L.M. Development of a Fabricable Gamma Prime (γ′) Strengthened Superalloy. *Superalloys* **2008**, *2008*, 191–200. [CrossRef]
6. Kruger, K.L. *15—HAYNES 282 alloy, Materials for Ultra-Supercritical and Advanced Ultra-Supercritical Power Plants*; Di Gianfrancesco, A., Ed.; Woodhead, Publishing: Sawston, UK, 2017; pp. 511–545. [CrossRef]
7. Jacobsson, J.; Andersson, J.; Brederholm, A.; Hänninen, H. Weldability of Ni-Based Superalloys Waspaloy® and Haynes® 282®: A Study Performed with Varestraint Testing. *Res. Rev. J. Mater. Sci.* **2016**, *4*, 3–11. [CrossRef]
8. Osoba, L.O. A Study on Laser Weldability Improvement of Newly Developed Haynes 282 Superalloy. Doctoral Thesis, University of Manitoba (Canada), Winnipeg, MB, Canada, 2012.
9. Caron., J.L. Weldability and welding metallurgy of HAYNES 282 alloy. In Proceedings of the 8th International Symposium on Superalloy 718 and Derivatives, Pittsburgh, PA, USA, 28 September–1 October 2014; John Wiley & Sons: Hoboken, NJ, USA, 2014. [CrossRef]
10. Hanning, F.; Andersson, J. Weldability of wrought Haynes® 282® repair welded using manual gas tungsten arc welding. *Weld. World* **2018**, *62*, 39–45. [CrossRef]
11. Caron, J.; Pike, L. Weldability of HAYNES 282 superalloy after long-term thermal exposure. In *MATEC Web of Conferences*; EDP Sciences: Les Ulis, France, 2014; Volume 14, p. 13003. [CrossRef]

12. White, H.; Santella, M.; Specht, E.D. Weldability of HAYNES 282 alloy for new fabrications and after service exposure. *Energy Mater.* **2009**, *4*, 84–91. [CrossRef]
13. Otto, R.; Brøtan, V.; Carvalho, P.A.; Reiersen, M.; Graff, J.S.; Sunding, M.F.; Berg, O.; Diplas, S.; Azar, A.S. Roadmap for additive manufacturing of HAYNES®282®superalloy by laser beam powder bed fusion (PBF-LB) technology. *Mater. Des.* **2021**, *204*, 109656. [CrossRef]
14. Unocic, K.; Kirka, M.; Cakmak, E.; Greeley, D.; Okello, A.; Dryepondt, S. Evaluation of additive electron beam melting of Haynes 282 alloy. *Mater. Sci. Eng. A* **2020**, *772*, 138607. [CrossRef]
15. Lim, B.; Chen, H.; Nomoto, K.; Chen, Z.; Saville, A.I.; Vogel, S.; Clarke, A.J.; Paradowska, A.; Reid, M.; Primig, S.; et al. Additively manufactured Haynes-282 monoliths containing thin wall struts of varying thicknesses. *Addit. Manuf.* **2022**, *59*, 103120. [CrossRef]
16. Shaikh, A.S.; Schulz, F.; Minet-Lallemand, K.; Hryha, E. Microstructure and mechanical properties of Haynes 282 superalloy produced by laser powder bed fusion. *Mater. Today Commun.* **2021**, *26*, 102038. [CrossRef]
17. Magnin, C.; Islam, Z.; Elbakhshwan, M.; Brittan, A.; Thoma, D.J.; Anderson, M.H. The performance of additively manufactured Haynes 282 in supercritical CO_2. *Mater. Sci. Eng. A* **2022**, *841*, 143007. [CrossRef]
18. Ramakrishnan, A.; Dinda, G.P. Microstructure and mechanical properties of direct laser metal deposited Haynes 282 superalloy. *Mater. Sci. Eng. A* **2019**, *748*, 347–356. [CrossRef]
19. Zhang, H.; Wang, Y.; De Vecchis, R.R.; Xiong, W. Evolution of carbide precipitates in Haynes® 282 superalloy processed by wire arc additive manufacturing. *J. Mater. Process. Technol.* **2022**, *305*, 117597. [CrossRef]
20. Xiong, W.; To, A. *Wire Arc Additive Manufacturing of Advanced Steam Cycle Components Using Location Specific Design Enhanced by High-Throughput Experiments and Machine Learning*; Physical Metallurgy and Materials Design Laboratory, University of Pittsburgh: Pittsburgh, PA, USA, 2020. Available online: https://netl.doe.gov/sites/default/files/netl-file/22FERD_HPM11_Xiong.pdf (accessed on 28 November 2022).
21. Selvi, S.; Vishvaksenan, A.; Rajasekar, E. Cold metal transfer (CMT) technology—An overview. *Def. Technol.* **2018**, *14*, 28–44. [CrossRef]
22. Wang, J.F.; Zhang, X.C.; Chuanchu, S. Microstructure and mechanical properties of Inconel 625 fabricated by wire-arc additive manufacturing. *Surf. Coatings Technol.* **2019**, *374*, 116–123. [CrossRef]
23. Kindermann, R.M.; Roy, M.; Morana, R.; Prangnell, P.B. Process response of Inconel 718 to wire + arc additive manufacturing with cold metal transfer. *Mater. Des.* **2020**, *195*, 109031. [CrossRef]
24. Xu, X.; Ding, J.; Ganguly, S.; Williams, S. Investigation of process factors affecting mechanical properties of INCONEL 718 superalloy in wire + arc additive manufacture process. *J. Mater. Process. Technol.* **2019**, *265*, 201–209. [CrossRef]
25. Zinke, M.; Burger, S.; Juettner, S. *Properties of Additively Manufactured Deposits of Alloy 718 Using CMT Process De-pending on Wire Batch and Shielding Gas*; IntechOpen: London, UK, 2022. [CrossRef]
26. Caron, J.L.; Sowards, J.W. *6.09—Weldability of Nickel-Base Alloys*; Hashmi, S., Batalha, G.F., Van Tyne, J.C., Yilbas, B., Eds.; Comprehensive Materials Processing; Elsevier: Amsterdam, The Netherlands, 2014; pp. 151–179. [CrossRef]
27. Jurić, I.; Garašić, I.; Bušić, M.; Kožuh, Z. Influence of Shielding Gas Composition on Structure and Mechanical Properties of Wire and Arc Additive Manufactured Inconel 625. *JOM* **2018**, *71*, 703–708. [CrossRef]
28. Agarwal, D.C.; Brill, U. High-temperature-strength nickel alloy. *Adv. Mater. Process.* **2000**, *158*, 31–34.
29. Herold, H.; Zinke, M.; Hübner, A. Investigations on the use of nitrogen shielding gas in welding and its influence on the hot crack behaviour of high-temperature resistant fully austenitic Ni- and Fe-base alloys. *Weld. World* **2005**, *49*, 50–63. [CrossRef]
30. Cutler, E.; Wasson, A.; Fuchs, G. Effect of minor alloying additions on the carbide morphology in a single crystal Ni-base superalloy. *Scr. Mater.* **2008**, *58*, 146–149. [CrossRef]
31. Huang, X.; Zhang, Y.; Liu, Y.; Hu, Z. Effect of small amount of nitrogen on carbide characteristics in unidirectional Ni-base superalloy. *Met. Mater. Trans. A* **1997**, *28*, 2143–2147. [CrossRef]
32. Nabavi, B.; Massoud, G. Nitrogen Effect on the Microstructure and Mechanical Properties of Nickel Alloys in Alloy 263 and Alloy. *Weld. J.* **2015**, *94*, 53–60. Available online: https://app.aws.org/wj/supplement/WJ_2015_02_s53.pdf (accessed on 28 November 2022).
33. Technical Datasheet AWS 062 Rev.1 HAYNES®282 of Alloy Wire International Ltd. Available online: https://www.alloywire.com/wp-content/uploads/2021/04/AW_Datasheet_HAYNES-282.pdf (accessed on 28 December 2022).
34. Wittig, B.; Zinke, M.; Jüttner, S.; Keil, D. Experimental simulation of dissimilar weld metal of high manganese steels by arc melting technique. *Weld. World* **2017**, *61*, 249–256. [CrossRef]
35. *ISO 14175:2008*; Welding Consumables—Gases and Gas Mixtures for Fusion Welding and Allied Processes. Beuth Verlag: GmbH, Berlin, 2008. [CrossRef]
36. *ISO 17636-1:2013*; Non-Destructive Testing of Welds—Radiographic Testing—Part 1: X- and Gamma-ray Techniques with Film. Beuth Verlag: GmbH, Berlin, 2013. [CrossRef]
37. *ISO 6892-1:2019*; Metallic Materials. Tensile Testing. Method of Test at Room Temperature. Beuth Verlag: GmbH, Berlin, 2019. [CrossRef]
38. *DIN 50125:2016-12*; Testing of Metallic Materials—Tensile Test Pieces. Beuth Verlag: GmbH, Berlin, 2016. [CrossRef]
39. *ISO 6507-1:2018*; Metallic Materials. Vickers Hardness Test. Test Method. Beuth Verlag: GmbH, Berlin, 2018. [CrossRef]
40. *ISO 6892-2:2018*; Metallic Materials—Tensile Testing—Part 2: Method of Test at Elevated Temperature. Beuth Verlag: GmbH, Berlin, 2018. [CrossRef]

41. Technical Datasheet HAYNES®282®ALLOY: A New, Wrought, Age-Hardenable, Nickel Superalloy Designed for Improved Creep Strength, Weldability, and Fabricability. Available online: http://www.haynes.ch/doc/haynes/282_h3173.pdf (accessed on 28 December 2022).
42. Hongfei, Y.; Yimi, G.; Yuefen, G. Effects of post-welding heat treatment on microstructure and mechanical properties of welding joint of new Ni-Fe based superalloy with Haynes 282 filler metal. *Mater. Sci. Eng. A* **2017**, *708*, 118–128. [CrossRef]

materials

Article

Optimization of CMT Characteristic Parameters for Swing Arc Additive Manufacturing of AZ91 Magnesium Alloy Based on Process Stability Analysis

Zhongrui Zhang [1,2], Junqi Shen [1,2,3,*], Shengsun Hu [1,2], Yang Chen [1,2], Chengxuan Yin [1,2] and Xianzheng Bu [4]

1 Tianjin Key Laboratory of Advanced Joining Technology, Tianjin University, Tianjin 300354, China
2 School of Materials Science and Engineering, Tianjin University, Tianjin 300354, China
3 International Institute for Innovative Design and Intelligent Manufacturing of Tianjin University, Shaoxing 312000, China
4 Beijing Hangxing Machinery Co., Ltd., Beijing 100013, China
* Correspondence: shenjunqi@tju.edu.cn; Tel.: +86-22-8535-6746

Abstract: The droplet transfer behavior and stability of the swing arc additive manufacturing process of AZ91 magnesium alloy based on the cold metal transfer (CMT) technique were studied by analyzing the electrical waveforms and high-speed droplet images as well as the forces on the droplet, and the Vilarinho regularity index for short-circuit transfer (IV_{SC}) based on variation coefficients was used to characterize the stability of the swing arc deposition process. The effect of the CMT characteristic parameters on the process stability was investigated; then, the optimization of the CMT characteristic parameters was realized based on the process stability analysis. The results show that the arc shape changed during the swing arc deposition process; thus, a horizontal component of the arc force was generated, which significantly affected the stability of the droplet transition. The burn phase current I_sc_wait presented a linear function relation with IV_{SC}, while the other three characteristic parameters, i.e., boost phase current I_boost, boost phase duration t_I_boost and short-circuiting current I_sc2, all had a quadratic correlation with IV_{SC}. A relation model of the CMT characteristic parameters and IV_{SC} was established based on the rotatable 3D central composite design; then, the optimization of the CMT characteristic parameters was realized using a multiple-response desirability function approach.

Keywords: magnesium alloy; swing arc additive manufacturing; CMT; droplet transfer; stability analysis; parameter optimization

Citation: Zhang, Z.; Shen, J.; Hu, S.; Chen, Y.; Yin, C.; Bu, X. Optimization of CMT Characteristic Parameters for Swing Arc Additive Manufacturing of AZ91 Magnesium Alloy Based on Process Stability Analysis. *Materials* **2023**, *16*, 3236. https://doi.org/10.3390/ma16083236

Academic Editors: Swarup Bag and Christ Prakash Paul

Received: 22 March 2023
Revised: 14 April 2023
Accepted: 18 April 2023
Published: 19 April 2023

1. Introduction

Metal resource consumption is increasing due to the needs of economic development, and most countries have made great efforts to strike a balance between economic growth and environmental protection. This pressure can be effectively reduced using high-performance lightweight metals, so their exploration and use have become popular research topics. As one of the Earth's most abundant metal elements, magnesium has a density of only 1.74 g/cm^3. Magnesium alloys have the characteristics of high specific strength and specific stiffness; excellent shock resistance and impact resistance; and good properties of thermal conductivity, damping and electromagnetic shielding. They also have the advantages of easy processing and recycling [1–4]. As a result, it has been discovered that magnesium alloys have significant application value and a wide range of potential applications in the fields of transportation, national defense, communication and medicine, and other industrial fields [5]. In engineering applications, casting is still the main processing method for magnesium alloys [6]. However, the tendency to obtain coarse grains when casting magnesium alloys results in poor mechanical properties and thus severely constrains the further application of magnesium alloys.

The direct production of large components and complex precision parts is greatly facilitated by additive manufacturing (AM). As one of the important techniques of AM, wire arc additive manufacturing (WAAM) is a direct energy deposition technique that uses arc welding procedures to fabricate components using layer-by-layer deposition, and it has been widely used for manufacturing parts of aluminum alloy [7,8], nickel-based alloy [9,10], stainless steel [11] and other metal materials. However, porosity, grain coarsening and thermal cracking are more likely to occur during the conventional WAAM process of magnesium alloys due to the high sensitivity to heat inputs induced by its unique physical and chemical properties [12]. As an improved technique of gas metal arc welding (GMAW), cold metal transfer (CMT) can realize a low-heat-input process with the precise control of the wire movement and the welding parameters by using a push–pull mechanism coupled with a high-speed digital process control system; thus, it helps to solve the above problems. Moreover, a stable and spatter-free WAAM process for magnesium alloys is made possible by the retraction control of the filler wire during the short-circuiting stage [13]. Therefore, the CMT technique shows significant technical advantages and broad application prospects in the fields of the WAAM of magnesium alloys.

By analyzing the electrical signals during the GMAW process, the arc stability can be determined as an index of the stability of the welding process [14]. Suban and Tusek [15] used the probability distribution of welding current to evaluate the influence of shielding gas components on arc stability during the GMAW process of low-carbon steel and found that the high current intensity in the short-circuit stage led to strong spatter caused by the breakage of the liquid bridge, making the welding process unstable. Chen et al. [16] explored the effect of the CMT characteristic parameters on the welding process stability of mild steel using the probability distributions of the CMT period and short-circuit time and indicated that fairly regular and stable processes could be achieved when the boost duration ranged from 1.6 ms to 3.6 ms at the boost current of 300 A. Singh et al. [17] used cyclograms of welding current vs. welding voltage to characterize arc stability and the spatter during the CMT weld-brazing process of aluminum/steel dissimilar alloys, and the results showed that the good repeatability of the cyclograms represented strong arc stability and the absence of spatter. The optimization of the welding parameters (i.e., welding current and welding voltage) by realizing the accurate control of the different stages of the GMAW process can reduce spatter and improve the stability of the welding process. Zhang et al. [18] achieved the desired one-droplet-per-pulse (ODPP) transfer by selecting the appropriate combination of the duration and amplitude of the peak current in the pulsed GMAW (GMAW-P) process of titanium alloy, and the stability of the GMAW-P process was greatly improved due to the precise control of the heat input in different stages of the droplet transition process. Subsequently, a much stronger droplet oscillation with a significantly lower heat input was realized with the active control of the current waveform during the GMAW-P process [19], which further increased the process stability.

Swing arc welding can effectively improve the weld appearance and grain refinement [20] and can also significantly reduce the formation of defects such as cracks and porosity [21,22]. In addition, the application of swing arc can increase the width of the deposited layer, thus improving the efficiency of WAAM [23]. However, much of the research on swing arc welding/deposition focuses on the microstructure and properties of the welded joint or deposited layer, while there is little analysis on the process stability when the swing arc is introduced. For example, though the spatter issue occurs when the swing arc is introduced [23], it has not attracted the attention of researchers. As is well known, welding spattering can lead to problems such as low wire utilization, poor bead formation, poor mechanical performance of the obtained joints and an extensive post-cleaning process [24–26]. In the multi-layer and multi-pass WAAM process, the spatter on the deposited pass is extremely detrimental to the deposition process stability of the subsequent pass and eventually to the performance of the WAAM component. Therefore, in order to ensure the appearance quality and performance of the deposited component, it is necessary to conduct research on the process stability of CMT swing arc deposition and the optimization of the CMT characteristic parameters.

As mentioned above, the introduction of the swing arc can cause spatter issues during the deposition process. Therefore, it is necessary to analyze the effect of the forces on the droplet during the CMT-WAAM process of AZ91 magnesium alloy based on the static equilibrium when the swing arc is applied, which can provide the theoretical analysis of spatter formation during the swing deposition process. Then, the influence of the CMT characteristic parameters in different stages on the stability of the swing arc deposition process can be investigated using the Vilarinho regularity index for short-circuit transfer (IV_{SC}) based on the coefficient of variation of electrical signals, as well as droplet transfer analysis. Finally, the stable swing arc deposition process can be achieved by optimizing the CMT characteristic parameters based on the above-mentioned process stability analysis.

2. Experimental Methodology

The substrate used in the experiment was an as-cast AZ91 magnesium alloy plate with dimensions of 200 mm × 150 mm × 8 mm, and the deposited material was AZ91 magnesium wire with a diameter of 1.2 mm. The chemical composition of the substrate plate and wire is illustrated in Table 1. The welding system was composed of a Motoman HP6 industrial robot and a Fronius Advanced 4000R welding machine equipped with a RCU 5000i controller, and the CMT program of C1904 was used to conduct the deposition process.

Table 1. Chemical composition of the substrate plate and wire (wt. %).

Material	Al	Zn	Mn	Si	Fe	Cu	Ni	Mg
Base metal plate	8.70	0.58	0.24	0.02	0.002	0.005	0.001	Bal.
Wire	8.62	0.55	0.33	0.0067	0.0040	0.00019	0.00023	Bal.

Before the deposition, the surface of the substrate plate was polished and wiped with ethanol to remove the oxide film and oils. The contact tip-to-work distance (CTWD) was maintained at 12 mm, and the welding gun was perpendicular to the substrate. Pure argon was used as the shielding gas, and the flow rate was 15 L/min. Based on the trial experiments and our previous research studies [27,28], the pre-flow time and post-flow time of the shielding gas were set at 1 s and 2.5 s, respectively, to ensure the protection effect on the start and end positions. During the deposition process, the substrate plate was fixed with a clamping device to prevent deformation. The introduction of the swing arc was realized using the 'single oscillation' weaving mode of the robot. The swing frequency and the swing amplitude were 5 Hz and 5 mm, respectively, and the dwell time on both sides and in the middle were both 0.1 s. The traveling speed of the welding torch was kept at 0.54 m/min.

Figure 1 shows the acquisition system consisting of a high-speed camera and an electrical signal acquisition device. The high-speed camera with an acquisition frequency of 500 fps was used to capture the droplet transfer, and a 1000 W high-voltage short-arc spherical xenon lamp was employed as the backlight source. A Hall current sensor and a Hall voltage sensor were used to capture the welding current and welding voltage signals, respectively, and a PCI-1742 data acquisition card with the sampling frequency of 50 kHz was used to send the electric signals to the computer. The program compiled using LabVIEW software was used to synchronously trigger the high-speed camera and electrical signal acquisition system.

A low-heat-input and spatter-free CMT welding process can be achieved as mentioned above, with the precise control of welding parameters such as welding current using high-speed digital process control. The schematic of the welding voltage and current waveforms used in the CMT welding of magnesium alloy is shown in Figure 2. There are four main characteristic parameters that can be controlled during the CMT welding process, i.e., boost phase current I_boost (A), boost phase duration t_I_boost (ms), burn phase current I_sc_wait (A), and short-circuiting current I_sc2 (A). When the arc is ignited after the detachment between the wire and the molten pool, the welding current is rapidly changed from I_sc2 to I_boost so as to keep the arc burning and the wire melting, and the duration of I_boost (i.e., the current of the peak phase) is controlled by t_I_boost. Then, the welding current is reduced to I_sc_wait, and the droplet gradually moves closer to the molten pool during the base phase. The short-circuit phase (S/C phase) occurs when the droplet comes into contact with the molten pool, and the welding current is immediately adjusted to the S/C pulse current and then is decreased to I_sc2. Finally, the droplet detaches from the wire into the molten pool, and a new CMT cycle occurs.

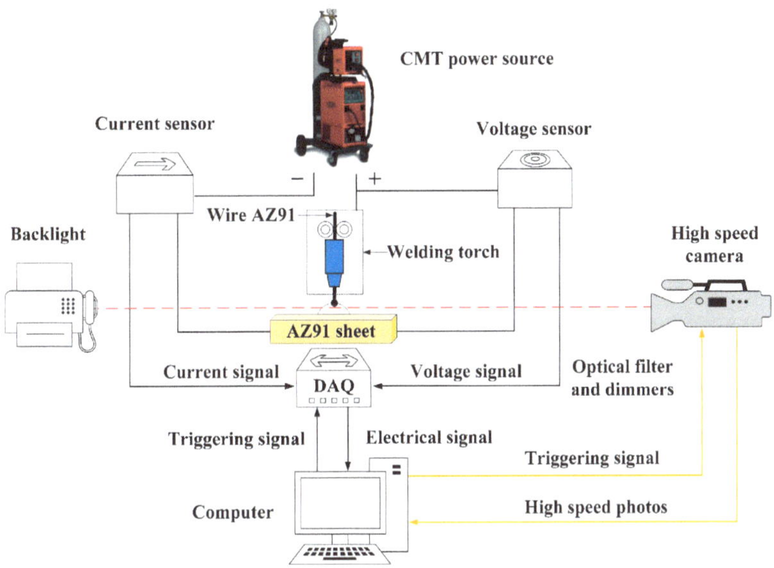

Figure 1. Schematic of the acquisition system.

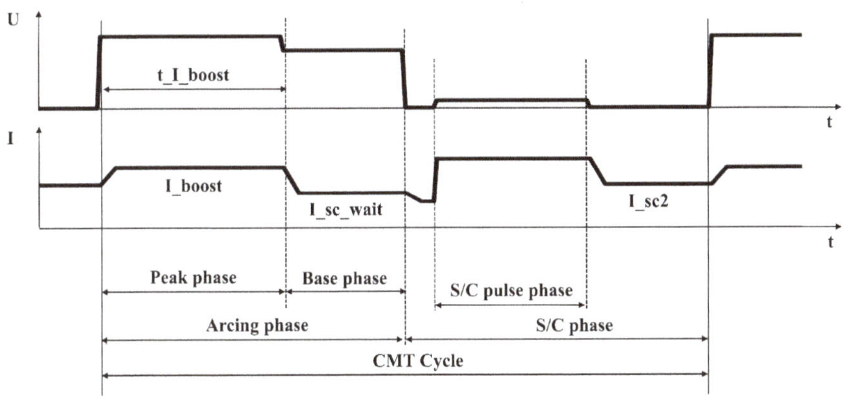

Figure 2. Schematic of voltage and current waveforms during CMT welding process of Mg alloy.

The precise control of the process stability in different stages can be achieved by adjusting the above CMT characteristic parameters. Based on the default electrical waveforms of the C1904 program at the wire feeding speed of 9 m/min (i.e., electric waveforms of sample number 1 in Table 2 as the basis for adjustment), the process stability under the conditions of different CMT characteristic parameters was analyzed. The CMT characteristic parameters for the swing arc deposition of Mg alloy are listed in Table 2.

Table 2. CMT characteristic parameters for the swing arc deposition of Mg alloy.

Sample Number	I_boost (A)	t_I_boost (ms)	I_sc_wait (A)	I_sc2 (A)
1	300	2	50	60
2	100	2	50	60
3	150	2	50	60
4	400	2	50	60
5	450	2	50	60
6	300	0.5	50	60
7	300	1	50	60
8	300	3	50	60
9	300	4	50	60
10	300	2	10	60
11	300	2	30	60
12	300	2	80	60
13	300	2	90	60
14	300	2	50	20
15	300	2	50	40
16	300	2	50	80
17	300	2	50	100

3. Results and Discussion

3.1. Cause Analysis of Spatter under Swing Arc

The forces on the droplet can be divided into two categories according to their effects on the droplet transition to the molten pool: positive ones and negative ones. The main forces acting on the droplet during the CMT deposition process include gravity (G), surface tension (F_γ), arc force (F_{arc}) and mechanical retraction force (F_b) [29]. The reason for the formation of spatter during the CMT swing arc deposition process of Mg alloy can be investigated by analyzing the forces on the droplet. Among the above forces, G is determined by the droplet quality (i.e., the amount of melted wire); F_γ is determined by the surface tension coefficient and the curvature radius of the liquid surface; F_{arc} is related to the current and the arc shape; and F_b mainly depends on the retraction speed of the wire. The introduction of the swing arc has little effect on the G, F_γ and F_b of the droplet, but it leads to an obvious change in F_{arc} due to the arc shape variation and thus affects the stability of the deposition process. During the CMT swing arc deposition process of Mg alloy, the arc force mainly includes electromagnetic force, plasma fluid force and reaction force of metal vapors.

Due to inertia, the arc and the droplet tend to keep moving in the same direction when the torch swings. Therefore, the arc shape becomes asymmetrical during the swing deposition process, as shown in Figure 3. The arc force during the swing deposition process is different from that during the linear deposition process, because the arc force is related to the arc shape [30]. Since the arc is bell-shaped and symmetrical along the wire axis during the linear deposition process, the arc force is also symmetrical, and there is no component in the horizontal direction. However, in the swing deposition process, the torch movement alters the shape of the arc and reduces its straightness, resulting in a horizontal component of the arc force acting on the droplet.

The electromagnetic force (F_m) is a vector with magnitude and direction, and it can be determined using Equation (1):

$$F_m = J \times B \tag{1}$$

where J is the current density and B is the electromagnetic flux density. During the linear deposition process, the current density is uniform, so the axial electromagnetic force is directed along the wire. However, during the swing arc deposition process, the offset arc acts on the side of the droplet, resulting in the uneven distribution of current density, as shown in Figure 4a. Therefore, the axial electromagnetic force hinders the droplet transition when the droplet diameter is greater than that of the wire. The axial electromagnetic force ($F_m{}^z$) on the droplet can be calculated according to Equation (2) [31]:

$$F_m{}^z = \frac{\mu_0 I_{arcing}{}^2}{4\pi}\left[ln\frac{r}{R}\sin\alpha - \frac{1}{4} - \frac{1}{1-\cos\alpha} + \frac{2}{(1-\cos\alpha)^2}ln\frac{2}{1+\cos\alpha}\right] \quad (2)$$

where μ_0 is the dielectric permeability, I_{arcing} is the current of the CMT arcing phase (i.e., I_boost and I_sc_wait) and α is the arc anode half angle.

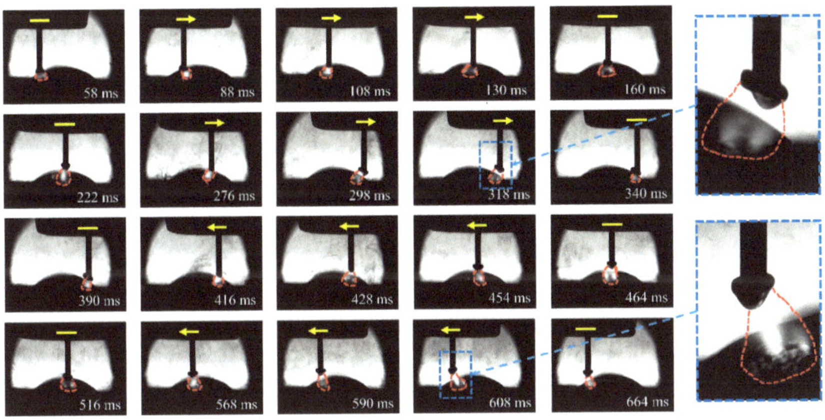

Figure 3. Arc shape variation during the swing deposition process.

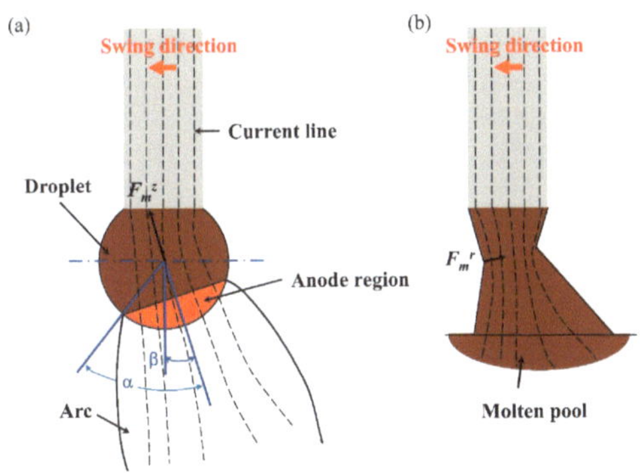

Figure 4. The electromagnetic force on the droplet: (**a**) formative stage; (**b**) transitional stage.

When the droplet comes into contact with the molten pool, the change in the droplet shape leads to a change in the current density, as shown in Figure 4b. At this moment, a radial electromagnetic force is generated at the necking, at which time the necking shrinks under the action of the radial electromagnetic force, thus promoting the droplet transition to the molten pool. The radial electromagnetic force ($F_m{}^r$) can be calculated according to Equation (3):

$$F_m{}^r = \frac{\mu_0 I_{sc}{}^2}{4\pi}\left(R^2 - r_0{}^2\right) \quad (3)$$

where I_{sc} is the current in the CMT short-circuit phase and r_0 is the radius of the necking.

The pressure near the electrode is greater than that near the substrate plate due to the uneven electromagnetic force inside the arc, and this pressure difference causes the particles in the arc to move towards the substrate, forming a flow pressure called plasma fluid force. The plasma fluid force (F_p) acting on the droplet can be calculated with Equation (4) [32]:

$$F_p = C_{D1} A_{p1} \left(\frac{\rho_p {v_p}^2}{2}\right) \tag{4}$$

where C_{D1} is the fluid drag coefficient, A_{p1} is the area of fluid action, ρ_p is the plasma density and v_p is the plasma movement velocity. As shown in Figure 5, the plasma fluid force acts along the arc axis, and the plasma fluid moves in the negative direction of the arc pressure gradient, thus promoting the droplet transition.

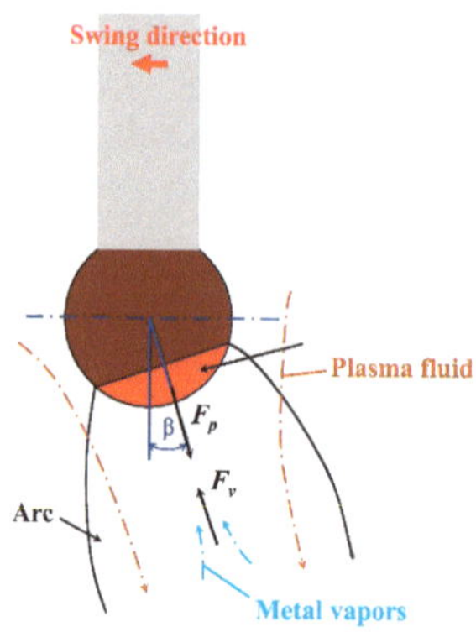

Figure 5. The plasma fluid force and reaction force of metal vapors on the droplet.

During the CMT deposition process, metal vapors are formed due to the low evaporation temperature of the magnesium alloy, and the reaction force of metal vapors pushes the droplet away from the arc axis [5]. The reaction force of the metal vapors (F_v) is also a kind of fluid force, and its magnitude and direction are determined by the flow and direction of the metal vapors. F_v can be calculated according to Equation (5) [32]:

$$F_v = C_{D2} A_{p2} \left(\frac{\rho_v {v_v}^2}{2}\right) \tag{5}$$

where C_{D2} is the metal vapor drag coefficient, A_{p2} is the area of metal vapor action, ρ_v is the metal vapor density and v_v is the metal vapor movement velocity.

The magnitude and direction of the arc force are related to the welding current and arc shape. During the CMT swing arc deposition process, the magnitude of the arc force does not change because the current remains stable. However, the change in the arc shape caused by the introduction of the swing arc alters the direction of the arc force and generates a component in the horizontal direction, which could lead to large spatter, especially during the multi-layer and multi-pass WAAM process. By adjusting the CMT characteristic parameters, it is achievable to control the forces acting on the droplet in order to eliminate the spatter and guarantee the stable droplet transition process.

3.2. *Effect of the CMT Characteristic Parameters on Process Stability*

The appearance of the deposited layer is shown in Figure 6. It is obvious that the aforementioned CMT characteristic parameters, i.e., boost phase current I_boost (A), boost phase duration t_I_boost (ms), burn phase current I_sc_wait (A) and short-circuiting current I_sc2 (A), had a significant impact on the appearance of the deposited layer due to their effects on the process stability.

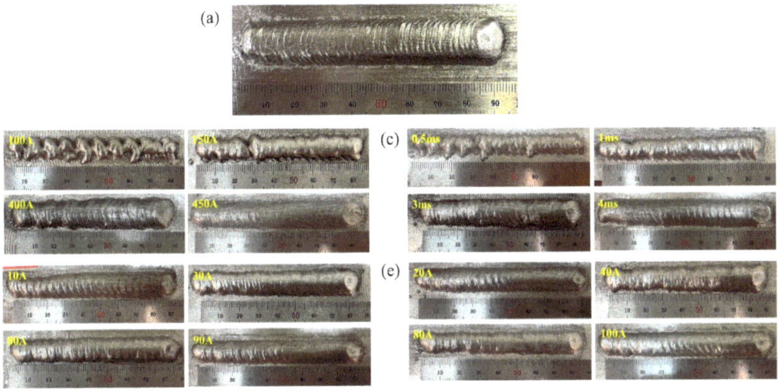

Figure 6. Appearance of the deposited layers: (**a**) Sample number 1; (**b**) Samples number 2–5; (**c**) Samples number 6–9; (**d**) Samples number 10–13; (**e**) Samples number 14–17.

The heat input (*HI*) of the CMT deposition process can be calculated according to Equation (6):

$$HI = \frac{\eta \sum_{i=1}^{n} UiIi}{nv_t} \quad (6)$$

where n is the number of electrical signal samples; U_i and I_i are the instantaneous voltage and instantaneous current, respectively; v_t is the traveling speed; and η is the thermal efficiency factor, which was taken as 0.7 in this paper. As shown in Table 3, the CMT characteristic parameters of t I_boost and t_I_boost had a significant effect on the heat input, while I_sc_wait and I_sc2 had little influence on the heat input.

Table 3. Phase duration, heat input and IV_{SC} with different CMT characteristic parameters.

Sample Number	Boots Phase (ms)	Burn Phase (ms)	S/C Phase (ms)	CMT Cycle (ms)	Heat Input (J/mm)	IV_{SC}
1	1.995	6.732	3.620	12.247	159.749	0.264
2	3.048	4.519	4.399	11.966	82.270	0.393
3	3.044	4.454	3.832	11.330	102.832	0.293
4	1.924	7.804	3.883	13.610	188.64	0.334
5	1.973	9.084	4.240	15.198	201.647	0.458
6	0.464	6.789	3.407	10.560	104.617	0.307
7	0.937	6.557	3.426	10.820	117.957	0.258
8	2.962	7.246	4.266	14.374	181.512	0.309
9	3.940	8.037	4.544	16.422	226.065	0.464
10	2.001	6.019	3.385	11.305	161.453	0.154
11	1.989	6.165	3.367	11.421	159.789	0.220
12	1.975	8.184	4.233	14.291	156.514	0.416
13	1.978	10.147	5.836	17.862	150.157	0.457
14	1.994	7.080	4.257	13.231	151.995	0.370
15	1.993	6.834	3.794	12.520	153.177	0.304
16	1.992	6.670	3.622	12.184	157.179	0.268
17	1.995	6.817	3.907	12.619	158.564	0.293

One way to characterize the process stability of CMT deposition is to analyze the stability of the electrical signals, and the common indicators for this analysis include the probability density distribution, *U-I* diagram, standard deviation and coefficient of variation. Probability density distribution and *U-I* diagram can provide a visual representation of the stability of the deposition process, but it is difficult to analyze them quantitatively. The coefficient of variation can not only provide the quantitative characterization of process stability for visual comparison but also eliminate the effects of standard deviation induced by the differences in measurement scales and magnitudes. Therefore, the coefficient of variation can provide a more accurate characterization of electrical signal dispersion. In CMT deposition, the process stability can be characterized using the Vilarinho regularity index (IV_{SC}) related to the coefficient of variation [33,34], which is calculated as shown in Equation (7):

$$IV_{SC} = \frac{\sigma_{tsc}}{t_{sc}} + \frac{\sigma_{tarcing}}{t_{arcing}} \tag{7}$$

where σ_{tsc} is the standard deviation of the short-circuit time, $\sigma_{tarcing}$ is the standard deviation of the arc burning time, t_{sc} is the mean short-circuit time and t_{arcing} is the mean arc burning time. As mentioned, because the regularity index (IV_{SC}) takes into account not only the standard deviation but also the mean value, a lower $IVsc$ indicates a more regular transition (i.e., stable process). Table 3 shows the IV_{SC} values with different CMT characteristic parameters.

It can be seen that the deposited layer was poorly formed and that the surface had clear 'grooves' at I_boost of 100 A (i.e., sample number 2) or t_I_boost of 0.5 ms (i.e., sample number 6), as indicated in Figure 6b,c. The unstable deposition process, as well as low heat input, resulted in a small amount of molten metal, poor wettability of the molten metal and discontinuous deposition process, eventually leading to a very poorly formed deposited layer.

The change in electrical signal stability can reflect the trend change of the droplet transition. Figure 7 shows the regression analysis results of the CMT characteristic parameters and coefficients. It is obvious that I_boost, t_I_boost and I_sc2 both displayed a quadratic correlation with IV_{SC}, while I_sc_wait displayed a good linear function relation with IV_{SC}.

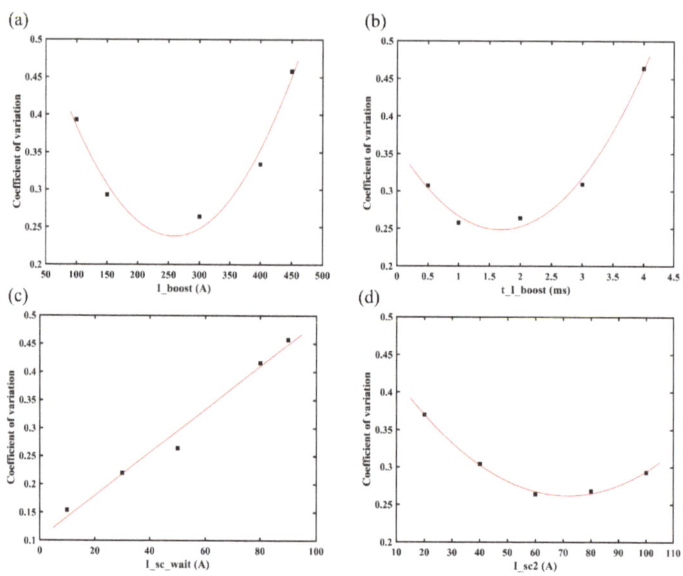

Figure 7. IV_{SC} of CMT deposition process at different (**a**) I_boost, (**b**) t_I_boost, (**c**) I_sc_wait and (**d**) I_sc2.

In order to clarify the effects of the CMT characteristic parameters on the droplet transition and process stability, the droplet transition process and the synchronous electrical signal waveforms were analyzed, and the high-speed droplet images corresponds to the area of electrical signal waveforms between the dotted lines. Figure 8 shows the electrical signal waveforms and the corresponding droplet transfer process at I_boost of 450 A (i.e., sample number 5) and t_I_boost of 4 ms (i.e., sample number 9). The electrical signal in both situations showed abnormal CMT cycles with very short duration, which manifested in an extremely short period in the base phase and the short-circuit phase. The corresponding droplet transition process showed that when the abnormal CMT cycle occurred, the droplet directly made contact with the molten pool to form a short-circuit phase and then rapidly detached from the wire. Therefore, the abnormal CMT cycle had no obvious base phase to maintain the arc burning nor short-circuit phase to facilitate the detachment of the droplet, thus leading to the instable deposition process. The high heat input of the peak phase not only melted more metal, but also reduced the viscosity coefficient of the liquid metal [35], so the droplet was prone to necking during the peak phase, and the radial electromagnetic force became a motive force for the droplet transition at that moment. Due to the rapid transfer of the droplet to the molten pool caused by the gravity and electromagnetic forces, the abnormal CMT cycle occurred. As shown in Figure 8a, when I_boost was 450 A, the high peak current led to a great instantaneous electromagnetic pinch force, resulting in a high probability of necking; thus, a greater number of abnormal CMT cycles appeared.

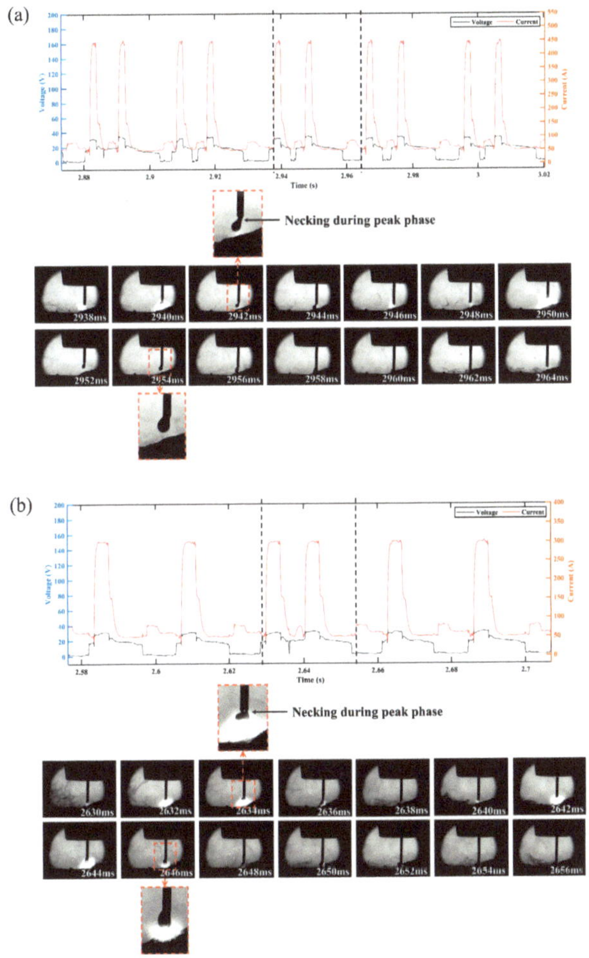

Figure 8. Electrical waveforms and droplet transfer process at (**a**) I_boost = 450 A and (**b**) t_I_boost = 4 ms.

The electrical signal waveforms and the corresponding droplet transfer process at I_sc_wait of 90 A (i.e., sample number 13) are shown in Figure 9. There were excessively long CMT cycles with abnormal short-circuit transition, leading to an increase in IV_{SC}.

The corresponding droplet transfer process showed that the droplet was spattered when it approached the molten pool, thus preventing the contact between the droplet and the molten pool. According to Equation (2), the excessive base current (i.e., high I_sc_wait) resulted in an increase in the arc force; thus, the repulsion of the droplet closer to the molten pool became more obvious, which not only resulted in a longer base phase but also caused spattering in an outward direction along the axis of the arc, thus destabilizing the CMT deposition process.

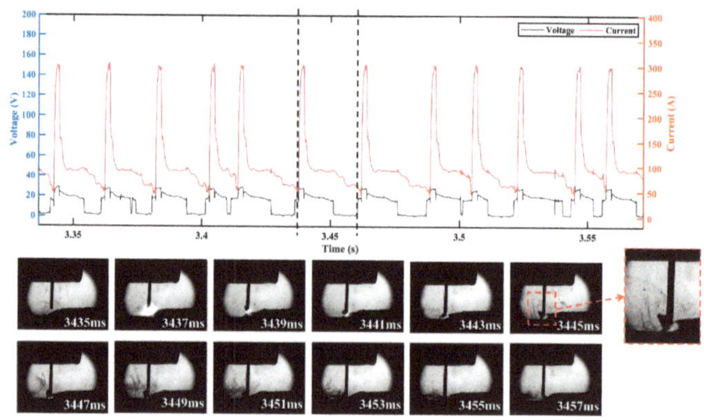

Figure 9. Electrical waveforms and droplet transfer process at I_sc_wait = 90 A.

Figure 10 shows the electrical signal waveforms and the corresponding droplet transfer process at I_sc2 of 20 A (i.e., sample number 14) and I_sc2 of 100 A (i.e., sample number 17). Under the condition of low or high short-circuit current, the liquid bridge formed at the end of the short-circuit phase became thick, making the liquid bridge prone to spattering when it was pulled off. However, the inducements and spattering patterns were different under these two conditions. When the short-circuit current was low, the spatter was located in an area where the liquid bridge was pulled off, as shown in Figure 10a, due to the high viscosity coefficient of the liquid metal induced by the low heat input of the short-circuit phase. When the short-circuit current was high, a relatively high current was passed through the necking, which caused the liquid bridge to explode due to overheating and vaporize, resulting in spattering. Therefore, the spatter was repelled in the outward direction with respect to the liquid bridge axis, as shown in Figure 10b.

3.3. *Optimization of the CMT Characteristic Parameters*

The process stability of the CMT swing arc deposition of AZ91 magnesium alloy is affected by the four above-mentioned CMT characteristic parameters, so the selection of CMT characteristic parameters for process optimization must be implemented using a proven and reliable method. Design of experiment (DoE) can provide a scientific/statistical approach for evaluating process variables, and it has the advantage of possible reduction in the number of experiments required for establishing relationships between input and output parameters [36]. In this paper, an experiment based on the rotatable 3D central composite design was used to systematically establish the relationship between CMT characteristic parameters and the corresponding IV_{SC}. For a set of measurable and controllable independent variables x_k, the response variable y can be expressed as

$$y = f\ (x_1,\ x_2,\ x_3,\ \ldots\ldots,\ x_k) \tag{8}$$

Based on the previous analysis, large I_sc_wait increased the repulsion, so the instability caused by I_sc_wait could be avoided by keeping it at a small value, i.e., I_sc_wait of 10 A. Therefore, I_boost, t_I_boost and I_sc2 were selected as the input parameters, with

IV_{SC} as the output parameter. The range of the CMT characteristic parameters for the swing arc deposition process are presented in Table 4.

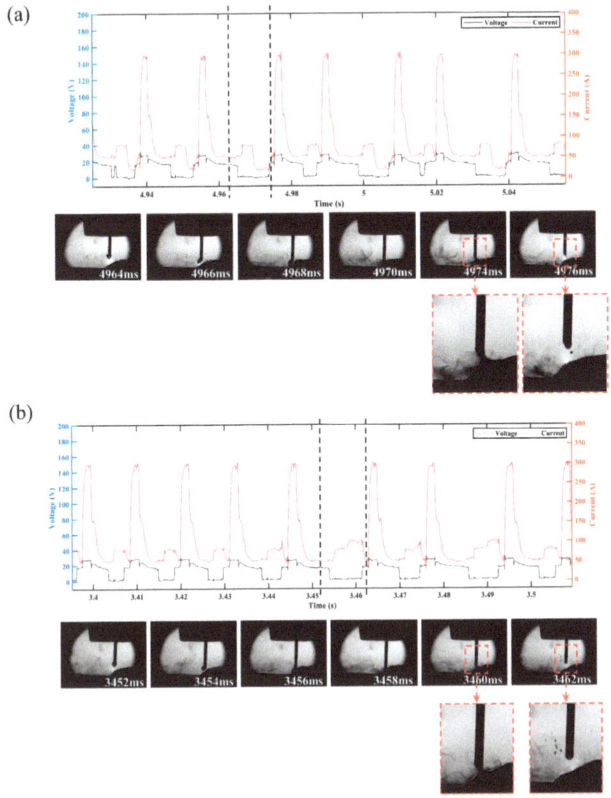

Figure 10. Electrical waveforms and droplet transfer process at (**a**) I_sc2 = 20 A and (**b**) I_sc2 = 100 A.

Table 4. Range of the CMT characteristic parameters.

Deposition Parameter	I_boost (A)	t_I_boost (ms)	I_sc2 (A)
Parameter range	170–330	1–2.5	50–90

As aforementioned, the design of experiment used in this paper was based on a rotatable central composite design, and the number of experimental runs n can be determined with Equation (9):

$$n = m_c + m_r + m_0 \tag{9}$$

where m_c is the number of two-level trials (corner points) and its value can be determined with $m_c = 2^p$; m_r is the number of trials at the asterisk (center points) and its value can be determined with $m_r = 2p$; m_0 is the number of zero-level trials (original point) and its value can be determined with $m_0 = r^4$; p is the number of input factors; and r is the length of the asterisk arm. As mentioned, the number of input factors p was 3, while the length of the asterisk arm r was 1.682 in the three-factor central composite design. Therefore, m_c, m_r and m_0 were 8, 6 and 9, respectively, and the number of experimental runs n was 23.

In this paper, the three experimental factors, i.e., I_boost, t_I_boost and I_sc2, were recorded as x_1, x_2 and x_3, respectively, and IV_{SC} was recorded as y. The upper and lower bounds of the jth factor are Z_{2j} and Z_{1j} (j = 1, 2, …, p), respectively, and the interval of zero-level variation Z_{0j} and the variation spacing Δ_j can be calculated according to Equations (10) and (11):

$$Z_{0j} = \frac{Z_{1j} + Z_{2j}}{2} \tag{10}$$

$$\Delta_j = \frac{Z_{2j} - Z_{0j}}{r} = \frac{Z_{2j} - Z_{1j}}{2r} \tag{11}$$

The true values of the corresponding level codes of I_boost, t_I_boost and I_sc2 were calculated according to Equations (11) and (12) respectively, and the results are shown in Table 5. The design table of the CMT characteristic parameters and the corresponding IV_{SC} are presented in Table 6. The CMT characteristic parameters for the swing arc deposition process were set based on a trial of a rotatable 3D central composite design.

Table 5. Variables and levels for the CMT characteristic parameters.

Level	I_boost (A)	t_I_boost (ms)	I_sc2 (A)
+1.682	330	2.5	90
+1	297.5	2.2	82
0	250	1.8	70
−1	202.5	1.3	58
−1.682	170	1	50

Table 6. Design table of the CMT characteristic parameters and the corresponding IV_{SC}.

Sample Number	Z_1	Z_2	Z_3	I_boost (A)	t_I_boost (ms)	I_sc2 (A)	IV_{SC}
1	1	1	1	297.5	2.2	82	0.187
2	1	1	−1	297.5	2.2	58	0.177
3	1	−1	1	297.5	1.3	82	0.410
4	1	−1	−1	297.5	1.3	58	0.452
5	−1	1	1	202.5	2.2	82	0.415
6	−1	1	−1	202.5	2.2	58	0.439
7	−1	−1	1	202.5	1.3	82	0.497
8	−1	−1	−1	202.5	1.3	58	0.461
9	1.682	0	0	330	1.8	70	0.226
10	−1.682	0	0	170	1.8	70	0.428
11	0	1.682	0	250	2.5	70	0.307
12	0	−1.682	0	250	1	70	0.482
13	0	0	1.682	250	1.8	90	0.349
14	0	0	−1.682	250	1.8	50	0.450
15	0	0	0	250	1.8	70	0.401
16	0	0	0	250	1.8	70	0.376
17	0	0	0	250	1.8	70	0.395
18	0	0	0	250	1.8	70	0.414
19	0	0	0	250	1.8	70	0.367
20	0	0	0	250	1.8	70	0.360
21	0	0	0	250	1.8	70	0.392
22	0	0	0	250	1.8	70	0.369
23	0	0	0	250	1.8	70	0.399

In the case of $p = 3$, mathematical models based on second-order polynomials were developed for predicting the responses. The models can be expressed as follows:

$$y = b_0 + \sum_{j=1}^{3} b_j x_j + \sum_{k=1}^{2} \sum_{j=2}^{3} b_{kj} x_k x_j + \sum_{j=1}^{3} b_{jj} x_j^2 \tag{12}$$

where x_1, x_2 and x_3 correspond to the three experimental input variables, i.e., I_boost, t_I_boost and I_sc2, respectively; y is the response variable, i.e., IV_{SC}; and b_0, b_j, b_k and b_{jj}

are the regression coefficients. Therefore, the mathematical model based on the results in Table 6 is as follows:

$$y = 0.39 - 0.068Z_1 - 0.066Z_2 - 0.014Z_3 - 0.049Z_1Z_2 - 0.0055Z_1Z_3 \\ -0.001Z_2Z_3 - 0.019Z_1^2 + 0.00446Z_2^2 + 0.006227Z_3^2 \tag{13}$$

Significance tests are required to ensure the credibility and fit of the regression equation. The analysis of variance for y is presented in Table 7. SS and df are the sum of squares and the degrees of freedom, respectively, and MS is the mean square calculated with SS/df. F indicates the significance, and a larger F represents a more significant variable and a better fit of the regression. P is the probability used to test the hypothesis at the 95% confidence level in this paper, and when its value is less than 0.05, it indicates that the factor is significant.

Table 7. Analysis of variance for response y.

Source	SS	df	MS	F	p	
Model	0.15	9	0.017	23.39	<0.0001	Significant
Z_1	0.063	1	0.063	87.60	<0.0001	Significant
Z_2	0.059	1	0.059	82.12	<0.0001	Significant
Z_3	0.002639	1	0.002639	3.68	0.0771	
Z_1Z_2	0.019	1	0.019	27.09	0.0002	Significant
Z_1Z_3	0.000242	1	0.000242	0.34	0.5710	
Z_2Z_3	0.000008	1	0.000008	0.011	0.9175	
${Z_1}^2$	0.005983	1	0.005983	8.35	0.0126	Significant
${Z_2}^2$	0.000316	1	0.000316	0.44	0.5182	
${Z_3}^2$	0.0006162	1	0.0006162	0.86	0.3706	

The results in Table 7 show that the F of the model is 23.39, indicating that the model is significant. For the condition of P less than 0.05, Z_1, Z_2, Z_1Z_2 and ${Z_1}^2$ are the terms with significant impact. Therefore, by removing the insignificant terms, the optimized regression equation can be obtained as follows:

$$y = 0.39 - 0.068Z_1 - 0.066Z_2 - 0.049Z_1Z_2 - 0.019Z_1^2 \tag{14}$$

The coefficient of determination R^2, as the percentage of the variability, indicates the degree of prediction of the regression model, and its maximum value is 1. The R^2 of Equation (14) is 0.8998, so the optimized regression equation has a high goodness of fit. By converting the above regression equation from the coded space to the real space, it can be expressed as follows:

$$y = -0.4591 + 0.00690518x_1 + 0.42933x_2 - 0.0023007x_1x_2 - 0.00000860779x_1^2 \tag{15}$$

The straight line in Figure 11 represents the predicted value of y (i.e., IV_{SC}) equal to its actual values. As shown in Figure 11, the actual values of y for the tests in Table 6 are distributed around the straight line, which can also indicate that the obtained regression equation fits significantly. Therefore, it can reflect the relationship between the CMT characteristic parameters and IV_{SC}.

As indicated by IV_{SC} with different CMT characteristic parameters, a regression model was established to predict the process stability of swing arc deposition. The results in Table 7 show that the effect of I_sc2 on the model is insignificant. The response surface of I_ boost, t_ I_boost and IV_{SC} at different I_sc2 levels within its range in Table 5 (i.e., 50–90 A) was established, and the results are shown in Figure 12. Using the response surface method, the experimental errors can be taken into account, and the complex functional relationships can be fitted to a continuous surface within a certain region. Thus, the optimal value of IV_{SC} with the lowest value can be found.

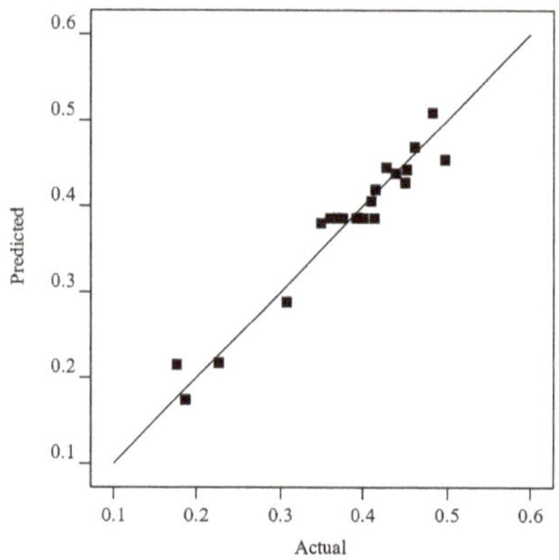

Figure 11. Predicted values versus actual values of IV_{SC}.

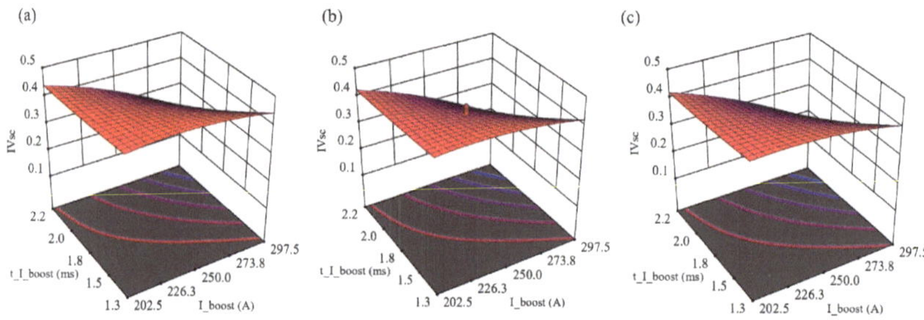

Figure 12. Response surface of I_ boost, t_ I_boost and IV_{SC} at different I_sc2: (**a**) I_sc2 = 58 A; (**b**) I_sc2 = 70 A; (**c**) I_sc2 = 82 A.

Table 8 shows the optimal combination of the CMT characteristic parameters (i.e., I_ boost, t_ I_boost and I_sc2) obtained with the established prediction model, as well as the corresponding predicted and actual IV_{SC}. Figure 13 shows the electrical signal waveforms and droplet transfer process of the deposited layer obtained using the optimum parameters in Table 8. As indicated in Figure 13, the electrical waveform was stable and free of abnormalities, and the phenomenon of droplet deviation during the CMT swing arc deposition process of Mg alloy could be eliminated using the optimized characteristic parameters.

Table 8. Optimization of the CMT characteristic parameters and the corresponding IV_{SC}.

Characteristic Parameters	I_boost (A)	t_I_boost (ms)	I_sc2 (A)	Predicted IV_{SC}	Actual IV_{SC}
Optimum	312.5	2.5	75.5	0.134	0.153

As shown in Figure 14, the generation of spatter could be effectively reduced during the multi-pass deposition process. Figure 15 shows that the deposited layer was composed of equiaxed grains, and there were significant differences in the average grain size in the transition zone between the adjacent deposited layers. The previous deposited layer could provide good heat dissipation conditions for the subsequent deposited layer, resulting in the formation of fine equiaxed grains on one side of the fusion line. Meanwhile, the previous deposited layer was reheated during the deposition process of the subsequent layer, leading to the formation of coarse equiaxed grains in the former layer near the fusion line. Furthermore, a defect-free microstructure was obtained, indicating good quality of the deposited layer and the interface of the successive deposited layers. In sum, a stable process of CMT swing arc deposition without spatter could be realized using the optimum parameters obtained with the established prediction model, and the deposited layers showed good appearance with no obvious defects.

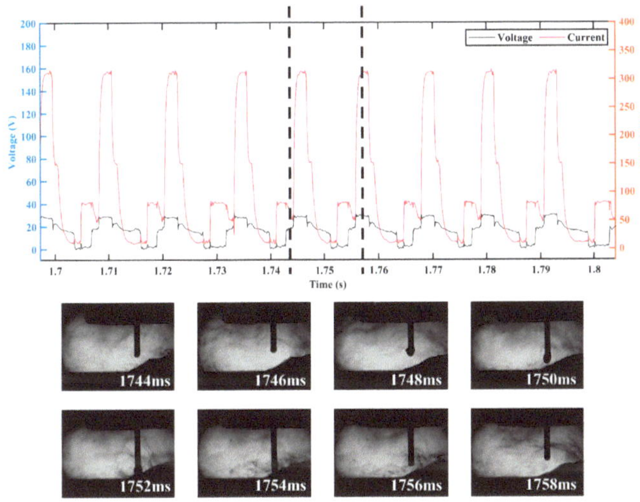

Figure 13. Electrical waveforms and droplet transfer process with the optimum CMT characteristic parameters.

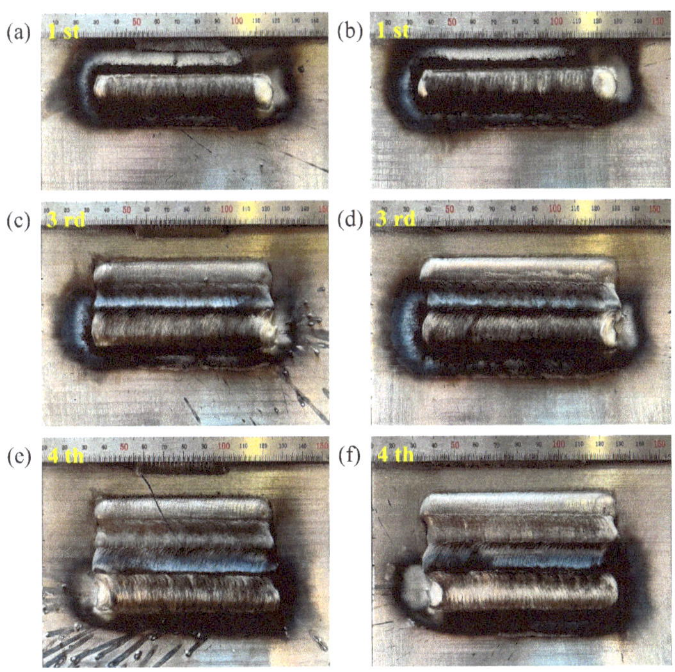

Figure 14. Appearance comparison: (**a**,**c**,**e**) 1, 3 and 4 deposited passes obtained before optimization, respectively; (**b**,**d**,**f**) 1, 3 and 4 deposited passes obtained after optimization, respectively.

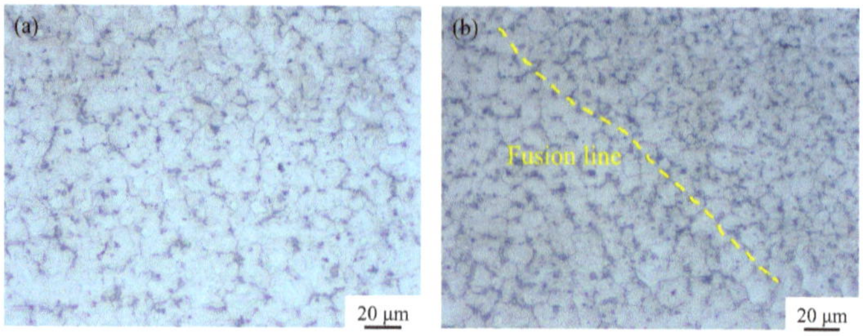

Figure 15. Microstructure of deposited layer: (**a**) interior of the deposited layer; (**b**) interface between successive deposited layers.

4. Conclusions

In this paper, a stable deposition process was realized by optimizing the CMT characteristic parameters, i.e., I_sc_wait, I_boost, t_I_boost and I_sc2. The primary conclusions obtained in this study are as follows:

(1) During the swing arc deposition process, a horizontal component of the arc force was formed, thus significantly affecting the stability of the droplet transition.
(2) The Vilarinho regularity index for short-circuit transfer (IV_{SC}) based on the coefficient of variation could be used to characterize the stability of the deposition process. I_sc_wait had a linear function relation with IV_{SC}, while I_boost, t_I_boost and I_sc2 all presented a quadratic correlation with IV_{SC}.
(3) A prediction model of IV_{SC} was established based on a rotatable 3D central composite design method; then, the optimization of the CMT characteristic parameters could be realized using this model.
(4) The optimum combination of CMT characteristic parameters for the swing arc deposition process of Mg alloy was I_boost of 312.5 A, t_I_boost of 2.5 ms, I_sc_wait of 10 A and I_sc2 of 75.5 A, which could guarantee a stable deposition process and good appearance of the defect-free deposited layer.

Author Contributions: Conceptualization, S.H.; Methodology, Y.C. and C.Y.; Formal analysis, Z.Z.; Investigation, C.Y. and X.B.; Resources, Y.C. and X.B.; Data curation, Z.Z.; Writing—original draft, Z.Z.; Writing—review & editing, J.S.; Supervision, J.S. and S.H.; Funding acquisition, J.S. All authors have read and agreed to the published version of the manuscript.

Funding: This research was funded by National Natural Science Foundation of China (grant No. 52075377).

Institutional Review Board Statement: Not applicable.

Informed Consent Statement: Not applicable.

Data Availability Statement: The raw/processed data required to reproduce these findings cannot be shared at this time due to technical or time limitations.

Conflicts of Interest: The authors declare that they have no known competing financial interests or personal relationships that could have appeared to influence the work reported in this paper.

References

1. Saremi, S.G.; Mirsalehi, S.E.; Shamsipur, A. Transient liquid phase bonding of AZ31 magnesium alloy: Metallurgical structure and mechanical properties. *J. Manuf. Process.* **2018**, *35*, 140–148. [CrossRef]
2. Li, W.Y.; Niu, P.L.; Yan, S.R.; Patel, V.; Wen, Q. Improving microstructural and tensile properties of AZ31B magnesium alloy joints by stationary shoulder friction stir welding. *J. Manuf. Process.* **2019**, *37*, 159–167. [CrossRef]
3. Kolodziejczak, P.; Kalita, W. Properties of CO_2 laser-welded butt joints of dissimilar magnesium alloys. *J. Mater. Process. Tech.* **2009**, *209*, 1122–1128. [CrossRef]

4. Song, G.; Diao, Z.; Lv, X.Z.; Liu, L.M. TIG and laser-TIG hybrid filler wire welding of casting and wrought dissimilar magnesium alloy. *J. Manuf. Process.* **2018**, *34*, 204–214. [CrossRef]
5. Hu, S.S.; Zhang, H.; Wang, Z.J.; Liang, Y.; Liu, Y.Q. The arc characteristics of cold metal transfer welding with AZ31 magnesium alloy wire. *J. Manuf. Process.* **2016**, *24*, 298–306. [CrossRef]
6. Weiler, J.P. Exploring the concept of castability in magnesium die-casting alloys. *J. Magnes. Alloys* **2021**, *9*, 102–111. [CrossRef]
7. Sun, J.M.; Hensel, J.; Köhler, M.; Dilger, K. Residual stress in wire and arc additively manufactured aluminum components. *J. Manuf. Process.* **2021**, *65*, 97–111. [CrossRef]
8. Vishnukumar, M.; Pramod, R.; Kannan, A.R. Wire arc additive manufacturing for repairing aluminium structures in marine applications. *Mater. Lett.* **2021**, *299*, 130112. [CrossRef]
9. Xi, N.Y.; Fang, X.W.; Duan, Y.S.; Zhang, Q.; Huang, K. Wire arc additive manufacturing of Inconel 718: Constitutive modelling and its microstructure basis. *J. Manuf. Process.* **2022**, *75*, 1134–1143. [CrossRef]
10. Seow, C.E.; Zhang, J.; Coules, H.E.; Wu, G.Y.; Jones, C.; Ding, J.L.; Williams, S. Effect of crack-like defects on the fracture behaviour of wire + arc additively manufactured nickel-base alloy 718. *Addit. Manuf.* **2020**, *36*, 101578. [CrossRef]
11. Zhang, Y.Q.; Cheng, F.J.; Wu, S.J. The microstructure and mechanical properties of duplex stainless steel components fabricated via flux-cored wire arc-additive manufacturing. *J. Manuf. Process.* **2021**, *69*, 204–214. [CrossRef]
12. Zhou, W.; Long, T.Z.; Mark, C.K. Hot cracking in tungsten inert gas welding of magnesium alloy AZ91D. *Mater. Sci. Tech.* **2007**, *23*, 1294–1299. [CrossRef]
13. Wang, P.; Zhang, H.Z.; Zhu, H.; Li, Q.Z.; Feng, M.N. Wire-arc additive manufacturing of AZ31 magnesium alloy fabricated by cold metal transfer heat source: Processing, microstructure, and mechanical behavior. *J. Mater. Process. Tech.* **2021**, *288*, 116895. [CrossRef]
14. Guo, N.; Xu, C.S.; Du, Y.P.; Chen, H.; Fu, Y.L.; Feng, J.C. Influence of calcium fluoride on underwater wet welding process stability. *Weld. World.* **2019**, *63*, 107–116. [CrossRef]
15. Suban, M.; Tušek, J. Methods for the determination of arc stability. *J. Mater. Process. Tech.* **2003**, *143–144*, 430–437. [CrossRef]
16. Chen, M.A.; Zhang, D.; Wu, C.S. Current waveform effects on CMT welding of mild steel. *J. Mater. Process. Tech.* **2017**, *243*, 395–404. [CrossRef]
17. Singh, J.; Arora, K.S.; Shukla, D.K. Lap weld-brazing of aluminium to steel using novel cold metal transfer process. *J. Mater. Process. Tech.* **2020**, *283*, 116728. [CrossRef]
18. Zhang, Y.M.; Li, P.J. Modified active control of metal transfer and pulsed GMAW of titanium. *Weld. J.* **2001**, *80*, 54–61.
19. Xiao, J.; Chen, S.J.; Zhang, G.J.; Zhang, Y.M. Effect of current waveform on droplet oscillation in GMAW. *Adv. Mat. Res.* **2013**, *652–654*, 2285–2288. [CrossRef]
20. Yuan, T.; Luo, Z.; Kou, S. Grain refining of magnesium welds by arc oscillation. *Acta. Mater.* **2016**, *116*, 166–176. [CrossRef]
21. Zhan, X.H.; Zhang, D.; Liu, X.B.; Chen, J.; Wei, Y.H.; Zhou, J.J.; Liu, R.P. Comparison between weave bead welding and multi-layer multi-pass welding for thick plate Invar steel. *Int. J. Adv. Manuf. Tech.* **2017**, *88*, 2211–2225. [CrossRef]
22. Zhu, C.X.; Cheon, J.; Tang, X.H.; Na, S.J.; Lu, F.G.; Cui, H.C. Effect of swing arc on molten pool behaviors in narrow-gap GMAW of 5083 Al-alloy. *J. Mater. Process. Tech.* **2018**, *259*, 243–258. [CrossRef]
23. Xu, X.F.; Ding, J.L.; Supriyo, G.; Diao, C.L.; Williams, S. Preliminary investigation of building strategies of maraging steel bulk material using wire+arc additive manufacture. *J. Mater. Eng. Perform.* **2018**, *28*, 594–600. [CrossRef]
24. Mai, S.; Kiyomitsu, K.; Rui, T.; Izumi, F.; Yu, Y.; Takayoshi, N.; Norimichi, T. Control system for reducing weld spatter during the welding process by using high-speed visual sensing. *Mech. Eng. Lett.* **2017**, *3*, 16–00679.
25. Liu, H.Y.; Li, Z.X.; Li, H.; Shi, Y.W. Study on metal transfer modes and welding spatter characteristics of self-shielded flux cored wire. *Sci. Technol. Weld. Join.* **2008**, *13*, 777–780. [CrossRef]
26. Molleda, F.; Mora, J.; Molleda, J.R.; Mora, E.; Mellor, B.G. The importance of spatter formed in shielded metal arc welding. *Mater. Charact.* **2007**, *58*, 936–940. [CrossRef]
27. Bi, J.; Shen, J.Q.; Hu, S.S.; Zhen, Y.H.; Yin, F.L.; Bu, X.Z. Microstructure and mechanical properties of AZ91 Mg alloy fabricated by cold metal transfer additive manufacturing. *Mater. Lett.* **2020**, *276*, 128185. [CrossRef]
28. Yin, C.X.; Shen, J.Q.; Hu, S.S.; Zhang, Z.R. Microstructure and mechanical properties of AZ91 magnesium alloy fabricated by multi-layer and multi-pass CMT based WAAM technique. *Result. Eng.* **2023**, *18*, 101065. [CrossRef]
29. Yan, Z.Y.; Zhao, Y.; Jiang, F.; Chen, S.J.; Li, F.; Cheng, W.; Ma, X.Q. Metal transfer behaviour of CMT-based step-over deposition in fabricating slant features. *J. Manuf. Process.* **2021**, *71*, 147–155. [CrossRef]
30. Yang, M.X.; Li, L.; Qi, B.J.; Zheng, H. Arc force and shapes with high-frequency pulsed-arc welding. *Sci. Technol. Weld. Join.* **2017**, *22*, 580–586. [CrossRef]
31. Fan, Y.Y.; Fan, C.L.; Yang, C.L. Research on short circuiting transfer mode of ultrasonic assisted GMAW method. *Sci. Technol. Weld. Join.* **2012**, *17*, 186–191. [CrossRef]
32. Kim, Y.S.; Eagar, T.W. Analysis of metal transfer in gas metal arc welding. *Weld. J.* **1993**, *72*, 269–278.
33. Meneses, V.A.D.; Gomes, J.F.P.; Scotti, A. The effect of metal transfer stability (spattering) on fume generation, morphology and composition in short-circuit MAG welding. *J. Mater. Process. Tech.* **2014**, *214*, 1388–1397. [CrossRef]
34. Truppel, G.H.; Angerhausen, M.; Pipinikas, A.; Reisgen, U.; Paes, L.E.D.S. Stability analysis of the cold metal transfer (CMT) brazing process for galvanized steel plates with $ZnAl_4$ filler metal. *Int. J. Adv. Manuf. Tech.* **2019**, *103*, 2485–2494. [CrossRef]

35. Song, K.Y.; Wang, Z.J.; Hu, S.S.; Zhang, S.Q.; Liang, E.B. Welding current influences on Inconel 625/X65 cladding interface. *Mater. Manuf. Process.* **2018**, *33*, 770–777. [CrossRef]
36. Shanmugam, R.; Ramoni, M.; Thangamani, G.; Thangaraj, M. Influence of additive manufactured stainless steel tool electrode on machinability of beta titanium alloy. *Metals* **2021**, *11*, 778. [CrossRef]

materials

MDPI

Article

Measuring the Cooling Behavior of Melt Pools in L-PBF by Pyrometry

Aron Pfaff [1,*], Sebastian Schäffer [1], Martin Jäcklein [1] and Frank Balle [1,2,3]

1 Fraunhofer Institute for High–Speed Dynamics, Ernst-Mach-Institut, Ernst-Zermelo-Str. 4, 79104 Freiburg, Germany; sebastian.schaeffer@emi.fraunhofer.de (S.S.); martin.jaecklein@emi.fraunhofer.de (M.J.); frank.balle@mail.inatech.uni-freiburg.de (F.B.)
2 Walter-and-Ingeborg-Herrmann Chair for Power Ultrasonics and Engineering of Functional Materials (EFM), Department for Sustainable Systems Engineering (INATECH), Faculty of Engineering, University of Freiburg, 79110 Freiburg, Germany
3 Freiburg Materials Research Center (FMF), Stefan-Meier-Str. 21, 79104 Freiburg, Germany
* Correspondence: aron.pfaff@emi.fraunhofer.de; Tel.: +49-761-2714-522

Abstract: This study aims to measure the cooling rates or, more precisely, the cooling durations of single laser tracks by pyrometry within the laser powder bed fusion (L-PBF) process. Two-color, as well as one-color pyrometers are tested within this work. Regarding the second, the emissivity of the investigated 30CrMoNb5-2 alloy is determined in-situ within the L-PBF system in order to measure temperature instead of arbitrary units. This is done by heating up printed samples and verifying the measured pyrometer signal by comparing it to values obtained by thermocouples attached to the samples. In addition, the precision of two-color pyrometry is verified for the given setup. Following the verification experiments, single laser track experiments are conducted. The obtained signals prove to be partially distorted mainly due to by-products such as smoke and weld beads arising from the melt pool. To encounter this problem, a new fitting method is presented and experimentally validated. Melt pools resulting from different cooling durations are analyzed by EBSD. These measurements show areas of extreme deformation or potential amorphization correlating with the cooling durations. The obtained cooling duration can be used for the validation of simulations as well as for the correlation of corresponding microstructure and process parameters.

Keywords: laser powder bed fusion; pyrometry; cooling behavior; 30CrMoNb5-2; cooling duration; microstructure

Citation: Pfaff, A.; Schäffer, S.; Jäcklein, M.; Balle, F. Measuring the Cooling Behavior of Melt Pools in L-PBF by Pyrometry. *Materials* **2023**, *16*, 3647. https://doi.org/10.3390/ma16103647

Academic Editors: Swarup Bag and Christ Prakash Paul

Received: 24 March 2023
Revised: 26 April 2023
Accepted: 5 May 2023
Published: 10 May 2023

1. Introduction

1.1. Motivation

Laser powder bed fusion (L-PBF) is the best-established metal additive manufacturing (AM) technology. Like most AM technologies, it is an intricate process due to the local creation of material [1]. Local imperfections or disturbances within the manufacturing process can cause defects. Therefore, a series of process monitoring technologies have been introduced. Co-axial pyrometry, as one of those, provides pointwise measurement, offering a high resolution as well as the possibility to measure the temperature within the melt pool. Thus, the technology is well suited to investigate cooling rates [2]. In general, the knowledge about cooling rates would support the validation of process simulations as well as the understanding of microstructural formation. Furthermore, the incremental nature of L-PBF offers the potential to apply locally varying process parameters in order to create functionally graded materials (FGMs), as shown by Popovich et al. [3], Pfaff et al. [4] or Kürnsteiner et al. [5]. Knowledge about the correlations between process parameters and cooling rates will support the design of further future FGMs.

1.2. State of the Art

McCann et al. [2], Everton et al. [6], Hou et al. [7] as well as Yang et al. [8] provide an overview of L-PBF process monitoring technologies and their challenges. Here, McCann et al. [2] point out the main problem of pyrometry: Without knowledge about the emissivity of the regarded material, only qualitative statements can be made. However, the emissivity depends on unknown factors such as temperature, surface condition and aggregate state. Mohr et al. [9] show a first approach to how emissivity can be measured within an L-PBF system. Pavlov et al. [10] point out another challenge regarding pyrometry. Depending on the measurement setup and the chosen process parameters, the observed melt pool can be smaller than the field of measurement. For one-channel pyrometry, however, the measured temperatures depend on the average intensity measured by the infra-red sensor resulting in an underestimation of the temperature. Gutknecht et al. [11] compare co-axial two-color pyrometry with thermographic and acoustic process monitoring. The measured temperatures seem to be below the melting point of the investigated material and are therefore presented as arbitrary units. Gutknecht et al. [11] are measuring co-axially to the laser beam and, therefore, within the melt pool. The change of aggregate state is named as the reason for arbitrary units, even though two-color pyrometry is used. In the field of laser welding, however, two-color pyrometry is used for quantitative measurements within the melt pool shown in the example by Xiao et al. [12]. Alternatives to pyrometer measurements are shown, for example, by Hooper [13] and Scipioni et al. [14], who measure melt pool cooling rates with high-speed cameras. Thermography, for example, presented by Altenburg et al. [15], is a popular approach to melt pool observations. However, the resolution is limited, and the measurement setup, as well as the data analysis, are complex. Farshidianfar et al. [16] use thermography in order to link the cooling rates with grain size, phase transformations and hardness. In addition, Takata et al. [17] link the cooling rates and resulting microstructures. Pyrometry measurements have not been linked yet with microstructural analysis. Based on Hooper [13], Scipioni et al. [14] and Farshidianfar et al. [16], cooling rates in the order of 10^1–10^7 K/s can be expected within L-PBF.

The literature describes pyrometry as a promising, affordable and robust candidate for melt pool recording as well as process monitoring in L-PBF (e.g., see McCann et al. [2], Everton et al. [6], Yang et al. [8], Gutknecht et al. [11] or Hooper [13]). Since most studies focus on the potential of using pyrometry for processing and quality monitoring (e.g., see [7,18–20]), the obtained signals are mainly interpreted as arbitrary units. Measuring temperatures instead of arbitrary units remains challenging due to the dependency of the emissivity from unknown factors such as temperature, surface condition and aggregate state.

1.3. Objective

The main goal of the presented research paper is to measure quantitative cooling rates by pyrometry. Therefore, an approach to measure quantitative temperatures within L-PBF instead of arbitrary units is introduced. Systematic single laser track experiments regarding the impact of laser power, scanning speed and laser focus diameter are combined with EBSD measurements in order to correlate the microstructures and cooling durations.

2. Materials and Methods

Within this work, the low-alloyed steel 30CrMoNb5-2 is investigated. The extended chemical composition of the alloy is listed in Table 1. Since the particle size distribution (PSD) is significantly impacting the absorptivity of laser radiation, the powder was characterized by a Camsizer X2 (producer: Microtrac Retsch GmbH), using high-speed imaging. The results are visualized in Figure 1, resulting in the following key parameters. The D10/50/90 values indicate that 10/50/90% of the particles have an even or smaller diameter than the presented value:

- D10 = 6.63 µm;
- D50 = 29.18 µm;
- D90 = 49.75 µm.

Table 1. Chemical composition of the used alloy in weight percent.

Element	C	Cr	Mn	Ni	Mo	Si	Cu	Al
wt. pct	0.29	0.91	0.73	0.33	0.27	0.26	0.05	0.04

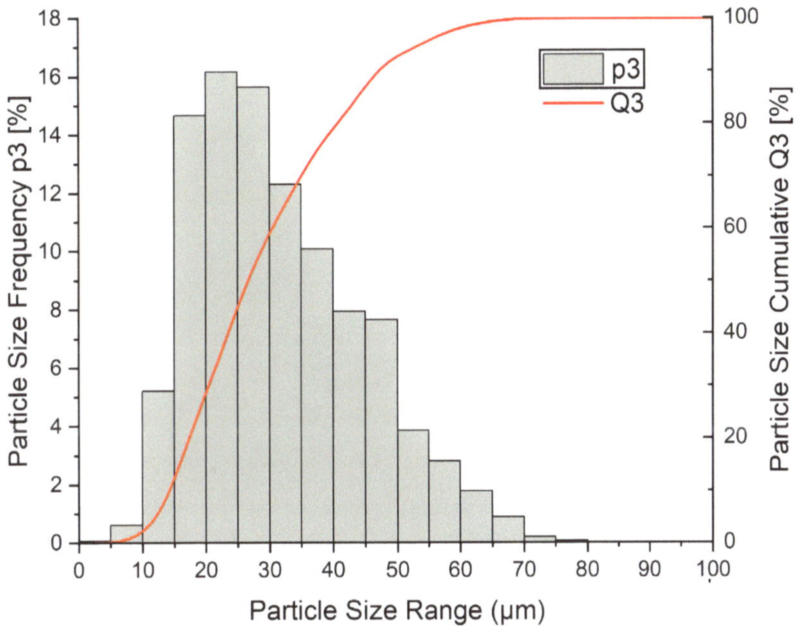

Figure 1. Particle size distribution of the used powder measured by analysis of high-speed imaging (Camsizer X2).

All pyrometer investigations were carried out using a Midi+ L-PBF system (producer: Aconity GmbH). Argon was utilized as an inert gas, resulting in an oxygen content of max. 400 ppm. The inert gas flow was set to 1 m/s in order to remove by-products of the welding process. The machine is equipped with four single-mode laser units (wavelength 1070 nm) and corresponding optical units. The optical path of the pyrometers is coaxially coupled into one of these units. This optical unit serves merely as a measurement tool, while the actual laser radiation is guided over a separate unit. The coaxial integration of the pyrometry is shown in Figure 2.

The scanners and the process chamber glass are water-cooled. An additional air cooling based on pressurized air can be activated for both components. According to the machine manufacturer, the setup results in a measurement field of ca. ⌀ 400 µm for the integrated pyrometry. The laser focus diameter can be varied between ca. 80–850 µm. All laser alignments, as well as the pyrometers, have been calibrated according to the manufacturer's instructions before the measurements.

Three one-color pyrometers at different wavelengths and one high-speed two-color pyrometer were used. Table 2 shows the minimum and maximum temperature as well as the type, frame rate, linearization status and measured wavelength for each infrared pyrometer. Furthermore, the table shows the notation used in the following graphs. The two non-linearized pyrometers, P3 and P4, can be combined into a two-color pyrometer. Pyrometers P3 and P4 are standard equipment installed by the machine producer, while pyrometers P1 and P2 were integrated by the authors. All data were recorded at 10 kHz using a transient recorder (MF-TransCom-CompaktX-XL). The fitting and data handling were executed in "Origin 2019b". The EBSD measurements were carried out using a DigiView 5 camera. The step width of the measurements was set to 0.1 μm.

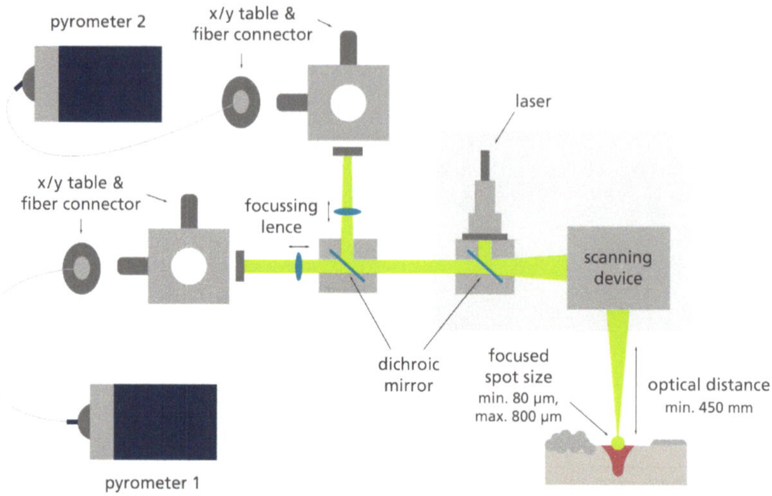

Figure 2. Optical path for the co-axial pyrometry within the used L-PBF system.

Table 2. Used IR pyrometry devices within this work and notations used in the following results.

	Kleiber KGA 740-LO	**METIS H322**	**Kleiber KGA 740-LO**	**Kleiber KGA 740-LO**
Notation	P1	P2	P3	P4
Type	one-color	two-color	one-color	one-color
Frame rate	100 kHz	25 kHz	100 kHz	100 kHz
Linearization	yes	yes	no	no
Tmin [°C]	200	400	500	500
Tmax [°C]	1000	1200	2500	2500
Wavelength [μm]	1.58–2.00	1.65–1.80 1.45–1.65	1.45–1.70	2.00–2.20

3. Results & Discussion

3.1. Realizing Quantitative Measurements by Pyrometry

3.1.1. Validation Setup

Figure 3 shows the experimental setup used to validate the temperature measured by two-color pyrometry as well as to determine the emissivity in dependency of the temperature. For temperatures up to 401 °C, a resistive heating system was used, and an inductive heating system was used for higher temperatures. The heating systems are separated by a base plate from the 2 mm thick substrate (here: C45 steel), which serves as a replaceable build platform. Two 30CrMoNb5-2 cuboid samples of 10 × 10 × 2 mm^3 are printed onto the substrate plate. A K-Type thermocouple with an accuracy of ±1 °K or ±0.75% is fixed to one of the samples using an M3 stainless steel screw. The thicknesses of the substrate and samples are minimized in order to avoid a loss of temperature between the heating system and the focus layer. When setting the machine to the maximum temperature of 1200 °C (measured inside of the machine close to the heating system), the thermocouple reached a maximum of 938 °C.

After reaching a stable temperature measured by the thermocouple, the samples are scanned by the pyrometers. One exemplary measurement result is visualized in Figure 4. The results show a slight deviation between the scanning field and the actual sample. This is evident by the increased pyrometer signal (yellow and orange signals) on the bottom side. The higher signals result from the radiation of the lower and, therefore, warmer substrate plate. Furthermore, the data recorded on the screwhead show slightly lower temperatures compared to the sample. The rest of the sample exhibits a homogenous signal. For the following investigation, one single scanning track close to the screw head and thermocouple was used in order to avoid tampered data outside the sample or on the screwhead. In general, the sample equipped with a screw and thermocouple shows a slightly lower temperature than the sample without a thermocouple. The difference is more significant for increasing temperatures. Thermocouples and screws seem to result in an increased heat transfer and a lower temperature due to convection. Since only the sample equipped with the thermocouple can be validated, the sample without is not further regarded within this work.

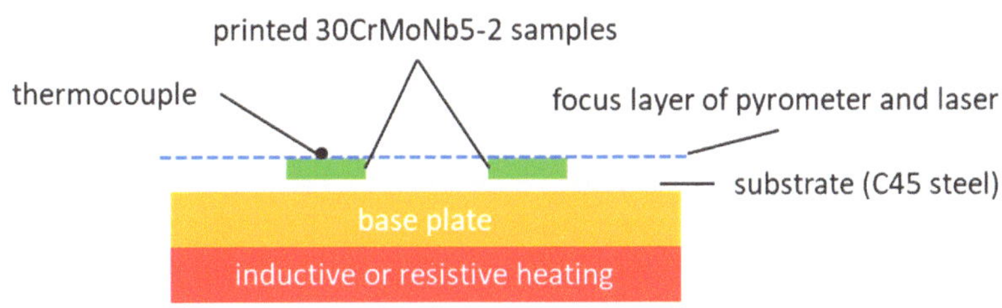

Figure 3. Experimental setup for the interpretation of the pyrometer signals.

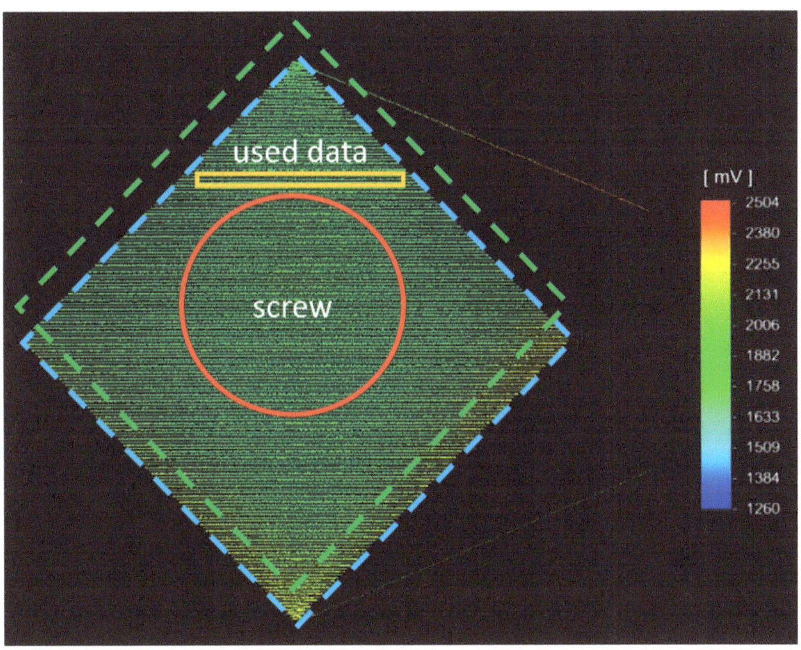

Figure 4. Data recorded while scanning the sample. Scanning field marked in blue. Sample marked in green. Screwhead marked in red. Relevant data used for the following analysis is marked in orange.

3.1.2. Determining the Emissivity by Linear Regression

In order to determine the emissivity by linear regression, measurements were carried out several times at each temperature while setting the emissivity value of the one-color pyrometers to levels between 0.1 and 1 (step size: 0.1). As long as the resulting temperature signal is lower than the temperature obtained by the thermocouple, the emissivity is set too high. The actual emissivity is determined by linear regression between the last measurement resulting in a too-low temperature and the following resulting in a too-high temperature. The resulting emissivities for all investigated temperatures are plotted in Figure 5.

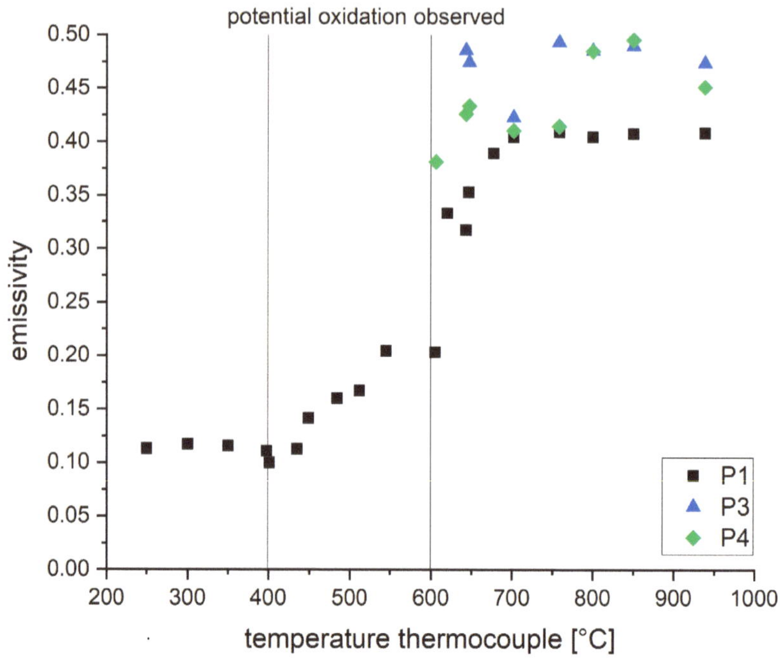

Figure 5. Temperature-dependent emissivity derived by linear regression for the one-color pyrometers P1 (wavelength 1.58–2.00 µm), P3 (wavelength 1.45–1.70 µm) and P4 (wavelength 2.00–2.20 µm).

Up to ~435 °C, the emissivity seems to be stable around an average of ~0.12, followed by a linear increase up to ~0.41. During this increase, a change in the color of the samples was observed. Furthermore, the oxygen level within the built chamber decreased from 400–600 ppm to around 150–300 ppm. It is, therefore, possible that oxides impacting the emissivity start to form at the surface of the samples. Between 646 °C and the highest measured value of 938 °C, the emissivity is constant around ~0.41. Regarding the non-linearized pyrometers, the determined emissivity within this last section is slightly higher (0.4–0.5). One aspect of this could be assigned to the fact that all three pyrometers measure different ranges of wavelength. Since the scanning device is not achromatically optimized, there could be optical aberrations generating an error in the measurement.

Thombansen et al. [21] describe the challenges of L-PBF processing in combination with a coaxially coupled process observation. Since optical elements such as f-theta lenses are monochromatically designed for the laser wavelength, chromatic aberrations of the thermal radiation could lead to a misinterpretation of temperature signals via pyrometry. In addition, the deviation could be caused by measurement noise in combination with the low significance within the calibration table for this temperature range. The calibration tables of P3 and P4 are visualized in Figure 6a. The curves show that temperatures up to 1000 °C result in a low signal of a maximum of ~4.3 mA (P3) or ~4.7 mA (P4). A typical signal, however, shows noise at a range of ±0.2 mA. A precise measurement is, therefore, not possible at these temperatures.

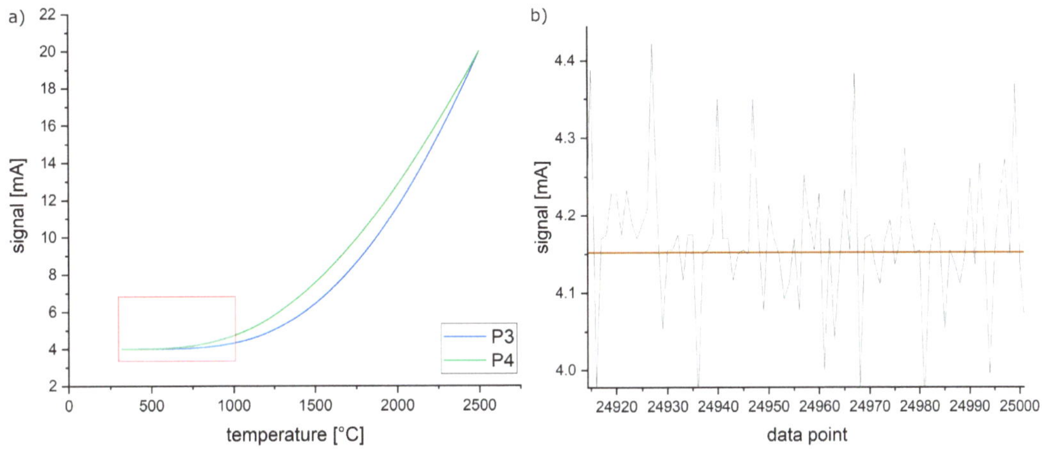

Figure 6. (**a**) Visualization of the manufacturer's calibration tables regarding the non-linearized pyrometers P3 and P4. Temperatures up to 1000 °C result in a low signal response (marked in red). (**b**) Signal obtained at approx. 700 °C shows pronounced noise, which results in a low signal-to-noise ratio (SNR). Red line represents the mean value used to correlate the corresponding temperature.

3.1.3. Determining the Emissivity by Comparison to a Theoretical Black Emitter

In general, the emissivity ε of a material is the ratio of the specific spectral radiation of a sample M_S and the theoretical specific radiation of the black emitter $M_{\Delta\lambda,}$ at the same temperature T_{Ref} and observed wavelength λ (Bernhard, 2014):

$$\varepsilon\left(T_{Ref}\right) = \frac{M_S\left(\Delta\lambda, T_{Ref}\right)}{M_{\Delta\lambda,}\left(\Delta\lambda, T_{Ref}\right)} \tag{1}$$

Therefore, an alternative approach to identify the emissivity of a surface with a known temperature is to measure the radiation at a set emissivity of 1 (assumption of a black emitter) and compare it to a theoretical black emitter:

$$\varepsilon(T) = \frac{\int_{\lambda_1}^{\lambda_2} M_S(\lambda, T_{\varepsilon=1})\mathrm{d}\lambda}{\int_{\lambda_1}^{\lambda_2} M_{\Delta\lambda}\left(\lambda, T_{Ref}\right)\mathrm{d}\lambda} = \frac{\displaystyle\int_{\lambda_1}^{\lambda_2} \frac{2\pi c_0^2 h}{\lambda^5\left(\mathrm{e}^{\frac{hc_0}{\lambda k T_{\varepsilon=1}}} - 1\right)}\mathrm{d}\lambda}{\displaystyle\int_{\lambda_1}^{\lambda_2} \frac{2\pi c_0^2 h}{\lambda^5\left(\mathrm{e}^{\frac{hc_0}{\lambda k T_{Ref}}} - 1\right)}\mathrm{d}\lambda} \tag{2}$$

h = Planck's quantum of action
k = Boltzmann constant
c_0 = light velocity in vacuum

This analytical approach enables the determination of the emissivity based on temperature signals derived from two independent sources (here: thermocouple and pyrometer). However, both integrals are evaluated over the same wavelength range. Therefore, two different black emitter spectra are regarded. Since the observed temperature signals from the thermocouple and pyrometer are close to each other, the absolute errors are assumed to be minor.

The emissivity calculated by this approach is shown in Figure 7. The results regarding P1 are in accordance with the alternative approach of linear regression. P3 and P4 show a lower deviation compared to the approach by linear regression. A pyrometer features a minimum error when adjusted to a maximum emissivity. The smaller fluctuations can therefore be associated with the assumed emissivity of 1 (see Formula (2)). In general, the approach presented by Formula (2) could also be used at lower assumed emissivity by considering the theoretical black emitter only partially. This approach has not been regarded since a lower assumed emissivity would also result in an increased error. However, due to the high setting of the emissivity, the resulting pyrometer signals are lower, resulting in the first suitable measurement signal above 850 °C.

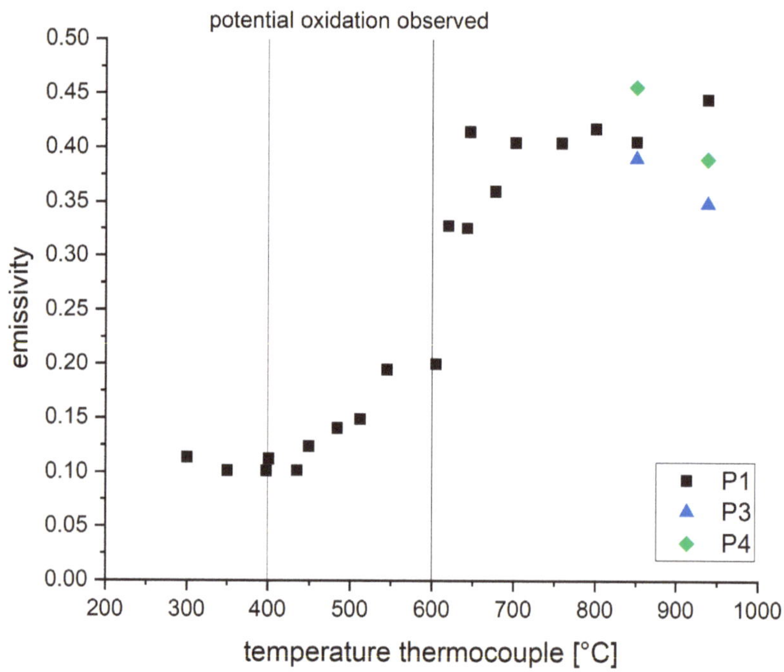

Figure 7. Temperature-dependent emissivity derived by comparison to a theoretical black emitter for the one-color pyrometers P1 (wavelength 1.58–2.00 μm), P3 (wavelength 1.45–1.70 μm) and P4 (wavelength 2.00–2.20 μm).

3.1.4. Validation of Two-Color Pyrometry

Two-color pyrometry takes advantage of Wien's approximation when putting Planck's radiation law at two different wavelengths in the ratio [22]:

$$\frac{I_1}{I_2} = \frac{\varepsilon_2}{\varepsilon_1}\left(\frac{\lambda_2}{\lambda_1}\right)^5 \frac{1 - exp\left(\frac{c_b}{\lambda_2 T}\right)}{1 - exp\left(\frac{c_b}{\lambda_1 T}\right)} \tag{3}$$

I = radiation intensities [$Wm^{-3}sr^{-1}$]

$c_b = 1.4388 \times 10^{-2}$ m K

And assuming that the emissivity is the same at both wavelengths (Wien's approximation):

$$\frac{I_1}{I_2} = \left(\frac{\lambda_2}{\lambda_1}\right)^5 \frac{1 - exp\left(\frac{c_b}{\lambda_2 T}\right)}{1 - exp\left(\frac{c_b}{\lambda_1 T}\right)} \tag{4}$$

The temperature T can be derived from the ratio of the intensities at different wavelengths:

$$T = \frac{c_b(\lambda_1 - \lambda_2)}{\lambda_1 \lambda_2} \frac{1}{\ln\left(\frac{I_1 \lambda_1^5}{I_2 \lambda_2^5}\right)} \tag{5}$$

However, Wien's approximation can cause certain measurement errors for certain materials, wavelengths, etc. Therefore, the two-color pyrometry approach is validated for the presented case. Figure 8 shows the deviation of the temperature obtained by two-color pyrometry for P2 and the combination of P3 and P4, compared to the temperature measured by the thermocouple. P2 shows a low deviation of ±2%.

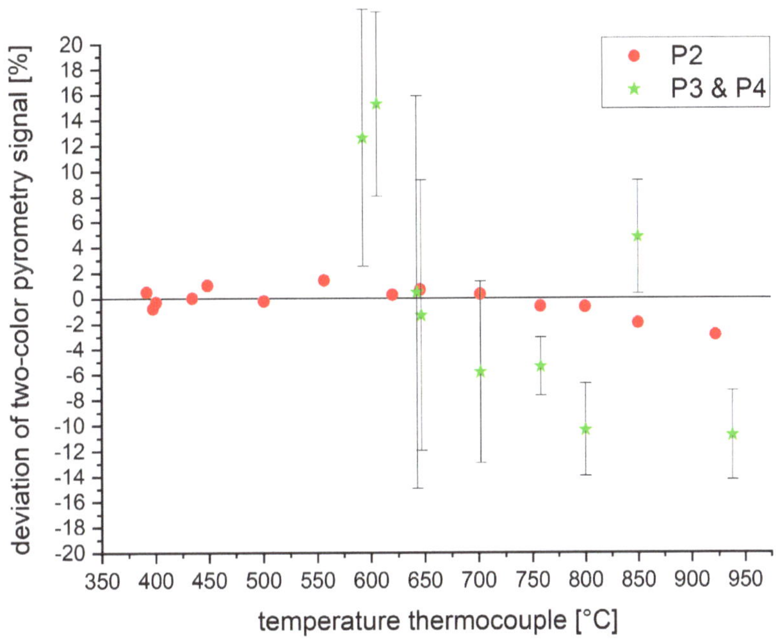

Figure 8. Validating the signals of the two-color pyrometers by comparing the measured temperatures to the signal of the thermocouple. Results are shown for the following pyrometers: P2 (wavelengths 1.45–1.65 and 1.65–1.80 μm) and the combination of P3 and P4 (wavelengths 1.45–1.70 and 2.00–2.20 μm).

The signals of P3 and P4 are combined by the formulas above. This is done for all measurements, which are executed at different emissivity levels between 0.1 and 1 (step size: 0.1). The results exhibit a higher absolute variation as well as significant standard deviations of the single measurements. Both could result from the investigated temperatures below 1000 °C, resulting in signals distorted by the measurement noise (see Figure 6).

3.2. *Approach to Measure Cooling Duration*

3.2.1. Experimental Setup

Figure 9 illustrates the measurement setup used in order to investigate the cooling rates or, more precisely, the cooling durations within this work. While the area observed by the pyrometer is resting statically in one position, a single laser track of defined laser power, scanning speed and focus diameter is passing through the field of measurement. The measurements are carried out on 30CrMoNb5-2 sample cubes manufactured by L-PBF measuring 50 × 50 × 10 mm^3. No powder layer is applied. In order to avoid any influence of inertia effects (e.g., starting movement of mirrors), the laser tracks measure a distance of 50 mm. The pyrometer is placed in the middle of this track. Each track is placed in a new position after the execution of an experiment in order to avoid the impacts of already exposed surfaces. Due to the noise effects regarding P3 and P4 observed in the validation process, the following investigations are conducted using P1 and P2. The emissivity of P1 is set to 0.12. Therefore, measurements above 435 °C will have to be adjusted based on the emissivity plotted in Figure 5 or Figure 7.

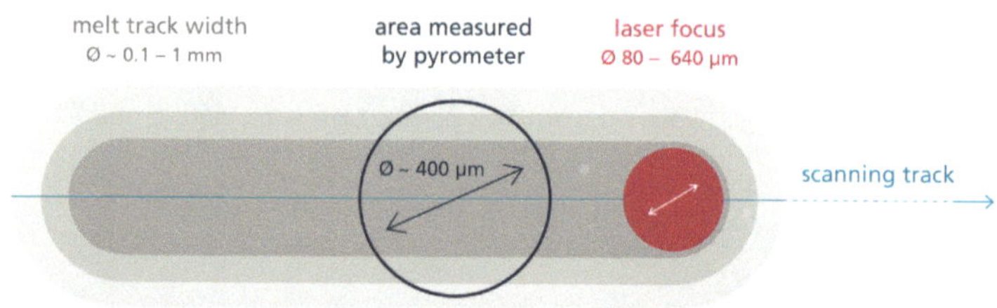

Figure 9. Measuring arrangement in order to characterize a single laser track of specific laser power and scanning speed.

It is known that the resulting melt track can be smaller than the area measured by the pyrometer. Regarding the one-color pyrometry, this can lead to a reduced signal since the average radiation within the field is being measured. The two-color pyrometry, however, should not be affected since only the ratio of the measured intensities at different wavelengths defines the measured temperature (see Formula (5)). Therefore, the maximum temperature within the area of measurement is given by the signal. According to Book [23], this is true as long as the maximum temperature is spreading over min. 20% of the measured area.

In order to ensure that the field of measurement is in alignment with the laser track, the XY-tables responsible for the positioning of the field of measurement (see Figure 2) have been moved until reaching the maximum signal within a single laser track experiment. This has been done additionally with the calibration according to the manufacturer's instructions.

3.2.2. Characteristics of Obtained Signals

Figure 10 shows characteristic signals obtained by the setup above. Measurements resulting in a smaller melt pool showed increased deviation between the one- and two-color pyrometers (for example, see Figure 10a). This can be explained by the fact that the measured area is bigger than the melt pool, resulting in a reduced signal. Bigger melt pools exhibit a smaller deviation (for example, see Figure 10b).

For some measurements, small signal peaks close to the actual peak were observed (see the brown mark in Figure 10b). This can be observed more often with increasing energy densities resulting in more by-products of the melt pool. It is therefore suspected that the peaks present welding beads flying through the field of measurement shortly before or after the passing of the actual laser focus.

In general, the signals show an asymmetric peak with a longer slope on the cooling side. The form of the peak is unexpected and does not correspond to simulation results, as shown by Karayagiz et al. [24,25]. An almost immediate increase followed by an exponential decline resulting in a sharp peak should be expected. Furthermore, the maximum temperature does not correlate with the melting temperature of the material, even though melt tracks are observed. The maximum temperatures increase with the applied energy density. This correlation, however, seems to be linear only in the beginning while converging around 700–750 °C when investigating a laser focus size of 80 µm. Several potential reasons for these unexpected signals are discussed in the following:

1. **Field of measurement not in line with melt track**

This option is ruled out since the maximum signal was obtained during the calibration process. Furthermore, there should be a pronounced deviation between one- and two-color pyrometry in all measurements (e.g., not the case in Figure 10b).

2. **Field of measurement is bigger than melt track**

This option is ruled out since parameter combinations causing melt pools close to 1 mm were investigated, resulting in the same type of signal. Furthermore, the use of two-color pyrometry should rule out this problem.

3. **Optical components warm up and distort signal**

This could have an impact. However, considering the short exposure time, the observed reduction of the temperature seems too high for this effect.

4. **By-products block the radiation from the melt pool**

This seems to be the most probable hypothesis. By-products were visible for almost all measurements. It would also explain why the temperature is converging with increasing energy densities. By-products rising from the melt pool would rapidly cool and start spreading after a short distance. Therefore, the resulting by-product plume is limited in its height and outside temperature. This effect should be more pronounced when exposing a powder layer since loose particles cause more by-products.

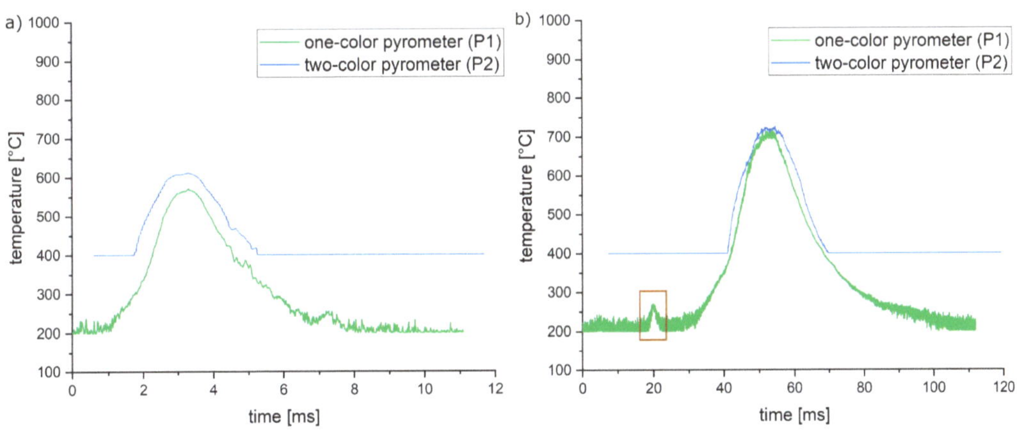

Figure 10. Obtained temperature signals of laser tracks showing typical melt pool by-products during exposure (e.g., smoke, welding beads, sparks). (**a**) Lower energy density: 400 W laser power, 2000 mm/s scanning speed and 80 µm focus diameter. (**b**) Highest investigated energy density: 900 W laser power, 250 mm/s scanning speed and 80 µm focus diameter. Brown mark: Potential welding bead flying through the measurement field of the pyrometer.

In order to check hypotheses three and four, process parameters resulting in no visible by-products when exposing bulk material but causing a plume of by-products when exposing a single powder layer are selected. A laser power of 400 W, a scanning speed of 1000 mm/s and a laser focus size of 640 µm are favorable for this verification approach. The parameters were applied under the following conditions:

1. Exposing a **powder layer** of 30 µm with bulk material below and **no air** cooling of the optical units. The black curve in Figure 11a.
2. Exposing **bulk material** directly **without air cooling** of the optical units. The red curve in Figure 11a.

3. Exposing **bulk material** directly and applying **additional air cooling** to the scanner unit and process chamber glass (set at 1.5 bars). Green (P1) and blue (P2) curves in Figure 11a.

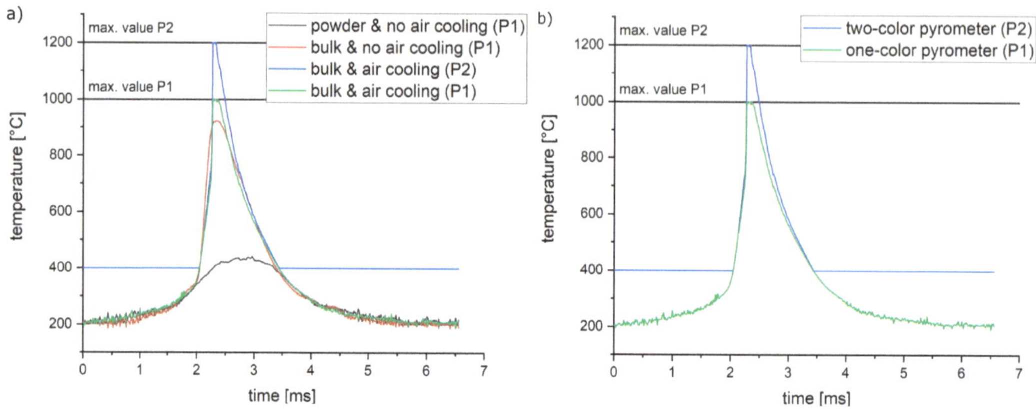

Figure 11. Comparing the impact of by-products and air-cooled optical components onto the temperature signal. Used laser track parameters: 400 W laser power, 1000 mm/s scanning speed and 640 μm focus diameter. (**a**) Comparing the exposure of a single 30 μm powder layer, using only water cooling (P1, black), the exposure of bulk material without air cooling (P1, red) and the P1 and P2 signals at the exposure of bulk material at additional air cooling of the optical components (green and blue). (**b**) Direct comparison of P1 and P2 at improved conditions (bulk material and air cooling).

The resulting temperature curves in Figure 11a show a strong impact on the by-products resulting from condition one (by-products caused by the powder layer). In addition, the use of additional air cooling of the optical units increases the maximum temperature. Using both optimized measurement conditions (no visible by-products and additional air cooling), the maximum temperatures of P1 and P2 are reached. Above 500 °C, Figure 11b also illustrates a small mismatch between the signals of P1 and P2. This deviation increases with the temperature and could be due to an increasing emissivity, as shown in Section 3.

The curves under optimized conditions exhibit the expectable behavior according to the literature (Karayagiz et al. [24,25]), showing a steep, rising slope followed by an exponential decline. Therefore, both hypotheses three and four prove to have an impact. However, by-products can only be avoided for a small amount of mainly irrelevant process parameter combinations. A methodology for signal processing of measurements distorted by by-products is presented in the following chapter.

3.2.3. Signal Processing for Systematic Investigations

The measurements in Figure 11 show that mainly the peak and the increase are distorted due to by-products. The declining slope, however, overlaps with the optimized signals after ~20% between the peak and the last valid signal. Therefore, the declining slope representing the cooling of the melt pool can be seen as valid and used to fit an analytical function (see Figure 12). It can be expected that the observed range of ~20% will only be valid for the regarded case since the size, speed and amount of by-products depend on the chosen process parameters and used material, as shown by Li et al. [26]. The presented validation experiment will therefore be necessary for each material or change in process parameters.

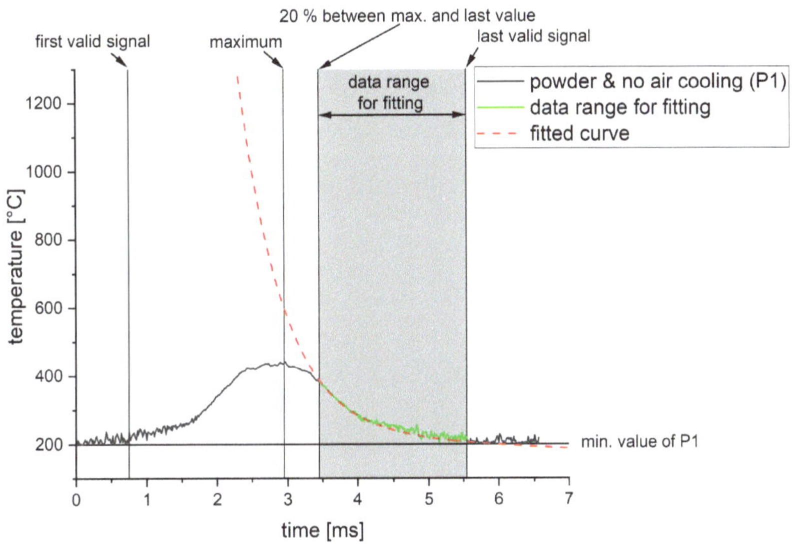

Figure 12. Definition of the data range used for fitting.

The melt pool's heat flow rate consists of conduction, convection and radiation:

$$\dot{Q} = \dot{Q}_{\kappa} + \dot{Q}_{\alpha} + \dot{Q}_{\varepsilon} \quad (6)$$

$$mc\frac{dT}{dt} = \kappa\frac{A_1}{d}(T_M - T) + \alpha A_1(T_M - T) + \varepsilon k A_2\left(T_M^4 - T^4\right) \quad (7)$$

A_1 = surface between melt pool and bulk material
A_2 = surface between melt pool and inert gas flow
κ = thermal conductivity
α = heat-transfer coefficient

The ambient temperature T is considered to be 0 °C. When replacing all constants with C_1, C_2, C_3 and C_4:

$$C_1\frac{dT_M}{dt} = C_2 T_M + C_3 T_M + C_4 T_M^4 \quad (8)$$

$$\int 1dt = \int \frac{C_1}{C_2 T_M + C_3 T_M + C_4 T_M^4} dT \quad (9)$$

And considering that C_1, C_2, C_3, C_4 and T_M are positive, the temperature of the melt pool T_M can be described by:

$$T_M = \frac{\sqrt[3]{-C_2 - C_3}}{\sqrt[3]{C_4 * -e^{-\frac{3Ct(C_2+C_3)}{C_1}}}} \quad (10)$$

When fitting this function to the data obtained by condition one (powder layer and no air cooling) and two (bulk material and no air cooling) within the range defined in Figure 12, the fitting curves in Figure 13 are obtained. Both fitting curves agree with the P2 signal under optimized conditions. The fitting curve for condition one (powder layer and no air cooling) shows no significant deviation. Therefore, the impact of a 30 μm thick powder layer on the cooling rate seems neglectable for the investigated parameter combinations. The presented approach is therefore used in all the following results.

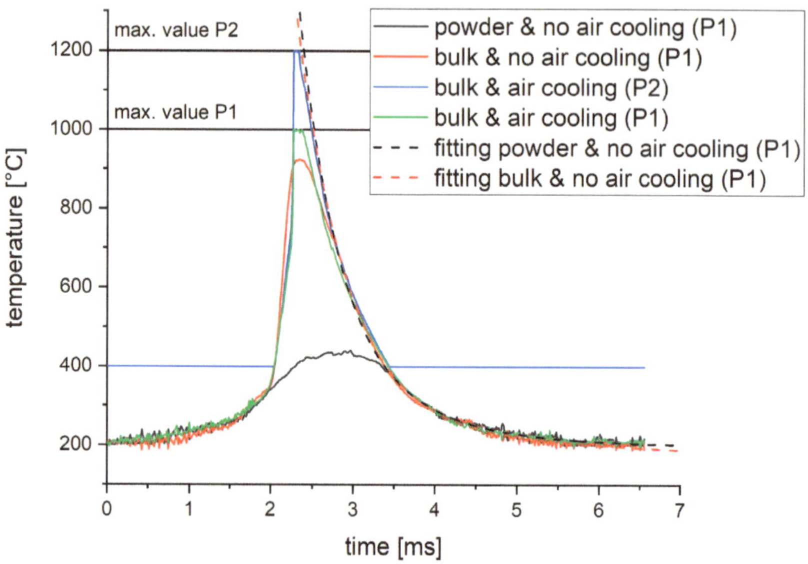

Figure 13. Comparison of the fitted curves and original temperature signals.

3.3. Measurements

3.3.1. Cooling Durations

The fact that the cooling rate [°C/s] presents a current value constantly changing over time (see different definitions of literature values in state of the art) makes it a rather unhandy parameter in order to predict microstructural evolutions. Therefore, the cooling duration between the austenitization temperature of around ~800 °C and typical values for the phase transformation to martensite and bainite are calculated based on the fitted curves. The resulting cooling duration between 800 °C and 500 °C (t8/5), 400 °C (t8/4), 300 °C (t8/3), as well as 200 °C (t8/2) are summarized in Figure 14. Cooling durations between ca. 0.5 and 70 ms are observed. The scanning speed has an exponential impact, while the laser power shows a linear effect on the cooling duration. Based on the coefficient of determination R^2 obtained by linear regressions, the linear correlation is especially significant for slower exposure speeds. At 2000 mm/s R^2 between 0.31 and 0.76, at 1000 mm/s R^2 between 0.62 and 0.92, at 500 mm/s R^2 between 0.86 and 0.93 and at 250 mm/s R^2 between 0.89 and 0.98 were observed. The reason for this could be unstable melt pools (balling phenomena) which occur and have been observed at higher exposure speeds. This would result in increased deviations within the measurements. Furthermore, the linear (impact of laser power) as well as the exponential regression (impact of exposure speed) show a higher significance for the cooling duration t8/5, t8/4 and t8/3 for t8/2. This could be due to the imprecision of the fitting process.

The impact of a changing laser focus diameter is shown in Figure 15. An increasing laser focus diameter results in a reduced cooling duration. One reason for this could be the changing geometry of the melt pool. The energy density decreases with increasing diameter. Therefore, melt pools will be wider and less deep, increasing the surface volume ratio and therefore increasing the heat flows. The impact of the laser focus diameter on the cooling duration is linear. Again, the significance reduces with increasing exposure speed, which could be caused by melt pool instabilities (see R^2 in Figure 15).

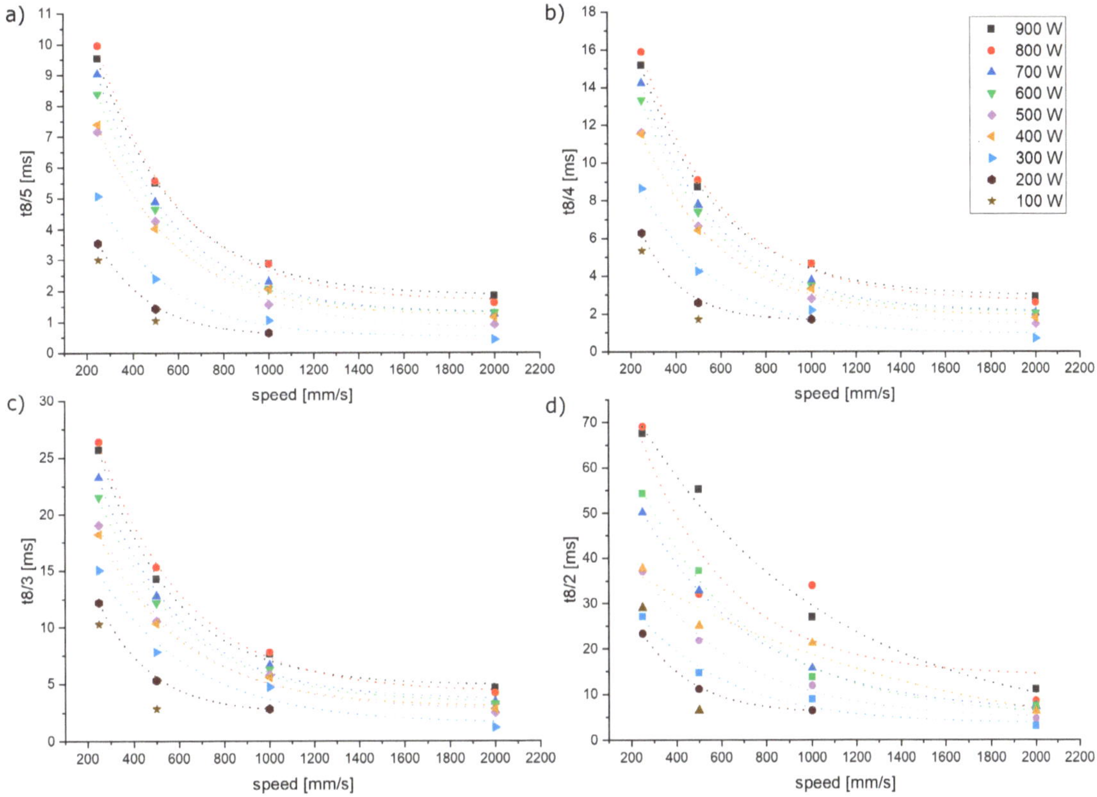

Figure 14. Derived cooling durations representing the cooling rates at a focus diameter of 80 µm. (**a**) 800 °C to 500 °C. (**b**) 800 °C to 400 °C. (**c**) 800 °C to 300 °C. (**d**) 800 °C to 200 °C.

A typical problem when setting up an L-PBF system for a print job is to position the coating system as precisely as possible within the laser focus layer. In order to investigate the impact of such an imprecision, the cooling durations were determined while giving the laser focus level an offset of up to 2 mm. The results (see Figure 16) do not show any significant correlation.

3.3.2. Microstructural Analysis

In order to investigate the impact of varying cooling duration on the resulting microstructures, EBSD measurements are carried out. This is done for single laser tracks in two different states:

1. **No heat treatment applied (no HT):** The laser tracks are placed within the final process layer. No following process layers are exposed.
2. **Heat treatment applied (HT):** The laser tracks are placed within the sample geometry. The following process layers are exposed by standard parameters. The exposure of each further layer results in an in-situ heat treatment of the single laser track.

Investigations by light microscopy on etched and non-etched samples did not show any significant characteristics. The EBSD results of the non-heat-treated single laser tracks are shown in Figure 17. All measurements for heat-treated and non-heat-treated single laser tracks reveal noise between areas of similar crystal orientation. The same phenomenon is observed within L-PBF bulk material. One reason for such a phenomenon could be an insufficient sample preparation resulting in surface-defects or -contamination. In order to rule out this problem, a sample manufactured by hot-rolling has been processed in the same way. No relevant amount of noise has been observed within this measurement. It is therefore concluded that the measured noise either represents areas of severe deformation or potential amorphization due to extremely high cooling rates. Luo et al. [27], for example, show similar EBSD results for amorphous metals manufactured by L-PBF. A higher resolution within the EBSD measurement could help to identify areas of severe deformation. However, a measurement of sufficient resolution was not successful yet.

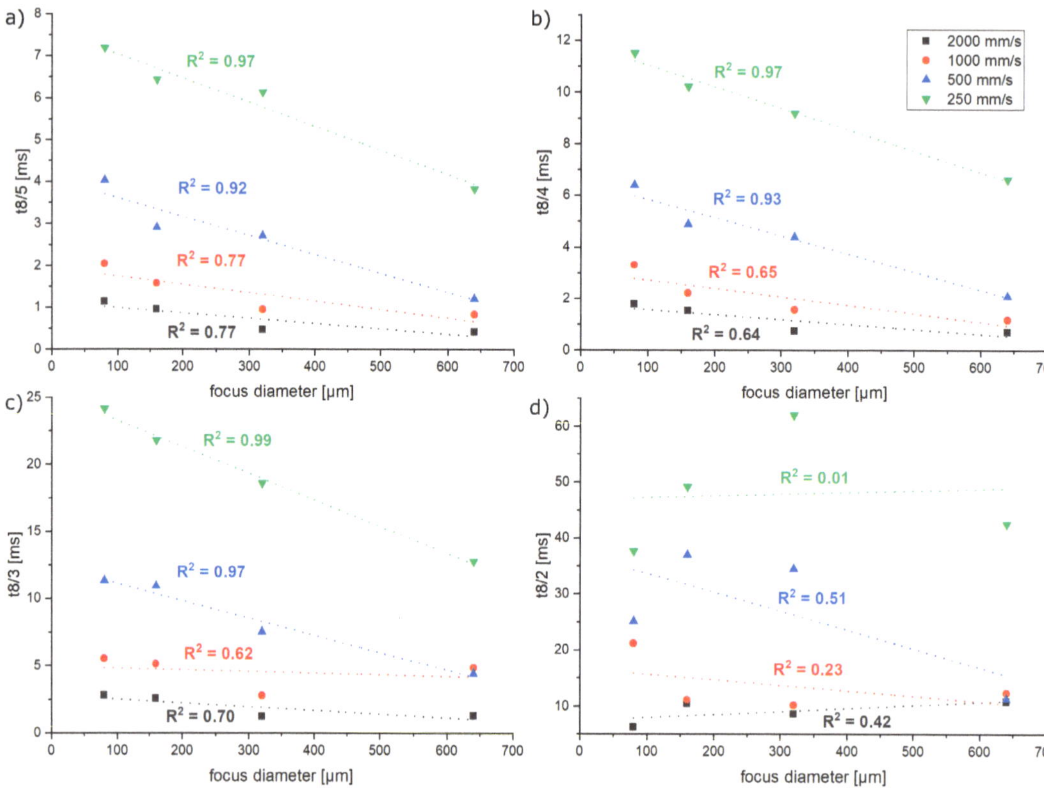

Figure 15. Impact of the focus diameter on the cooling durations. (**a**) 800 °C to 500 °C. (**b**) 800 °C to 400 °C. (**c**) 800 °C to 300 °C. (**d**) 800 °C to 200 °C.

When comparing the fraction of noise within the single laser tracks, a clear decreasing trend with increasing cooling duration is visible. Figure 18 shows the correlation between the cooling duration t8/5 and the resulting ratio of noise. This trend is in accordance with the assumption of microstructures resulting from severe deformations or amorphous areas. While the in-situ heat treatment of the laser tracks does not show any significant effect on the fraction of noise regarding slower cooling durations of 24.5 ms and 10.9 ms, the laser track resulting from a cooling duration of 6.9 ms shows a significant decrease in noise. This could be due to annealing effects.

It is well known that the grain size within a microstructure will increase with the cooling duration. Therefore, Figure 19 presents the average grain size in μm^2 for the investigated cooling durations. The results show a tendency to increase grain sizes. The in-situ heat treatment does not show any significant impact on the grain size.

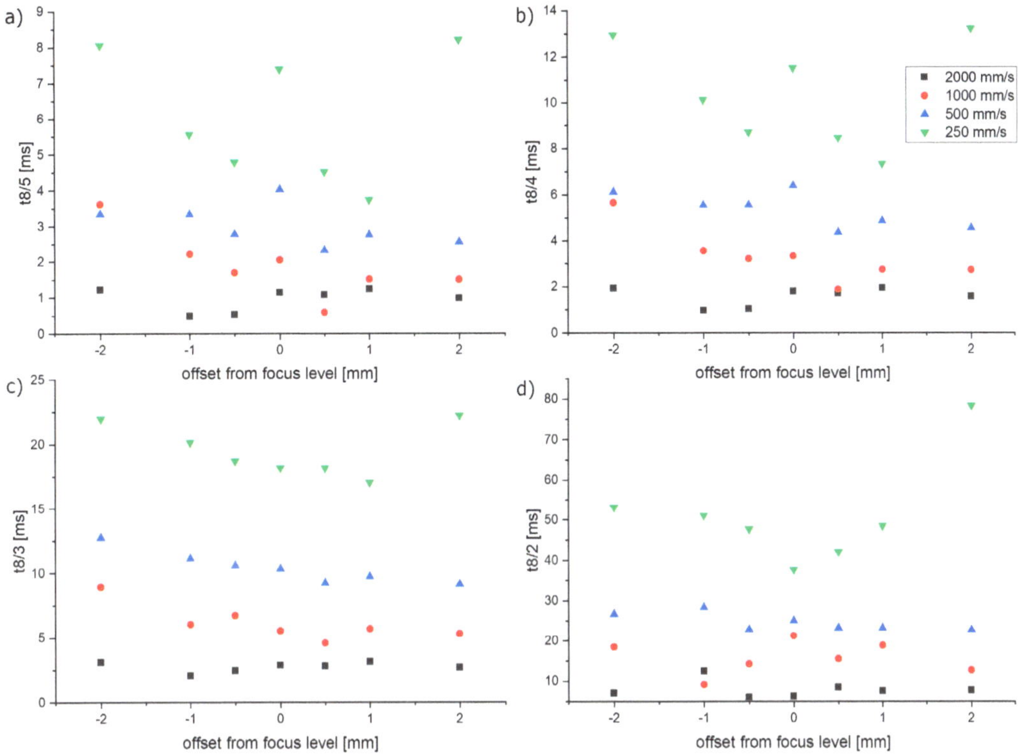

Figure 16. Impact of the focus level offset on the cooling durations. (**a**) 800 °C to 500 °C. (**b**) 800 °C to 400 °C. (**c**) 800 °C to 300 °C. (**d**) 800 °C to 200 °C.

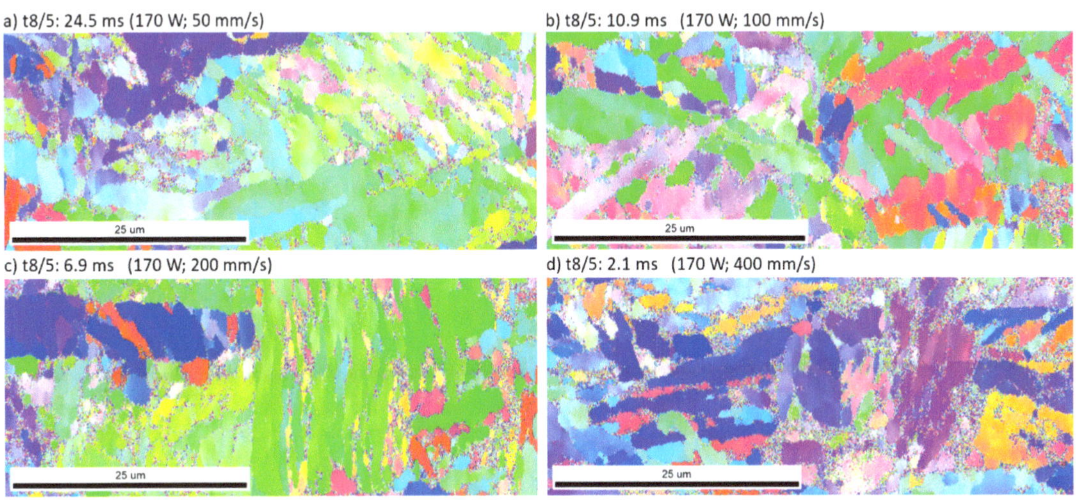

Figure 17. *Cont.*

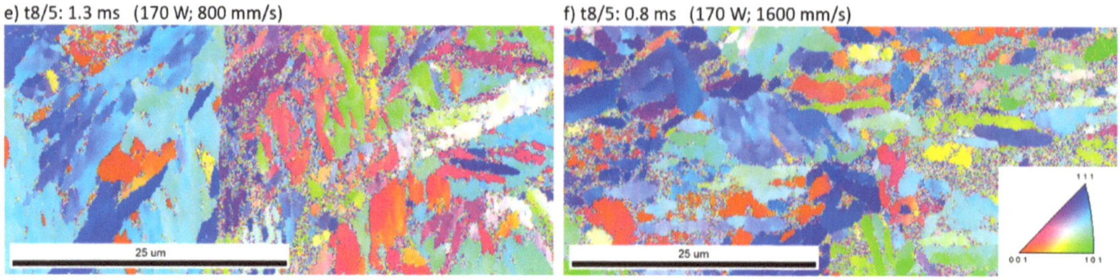

Figure 17. EBSD measurements at non-heat-treated single laser tracks (laser tracks within the last process layer). The cooling duration t8/5 is varied by increasing exposure speeds and a constant laser power of 170 W. (**a**) 24.5 ms (50 mm/s); (**b**) 10.9 ms (100 mm/s); (**c**) 6.9 ms (200 mm/s); (**d**) 2.1 ms (400 mm/s); (**e**) 1.3 ms (800 mm/s); (**f**) 0.8 ms (1600 mm/s).

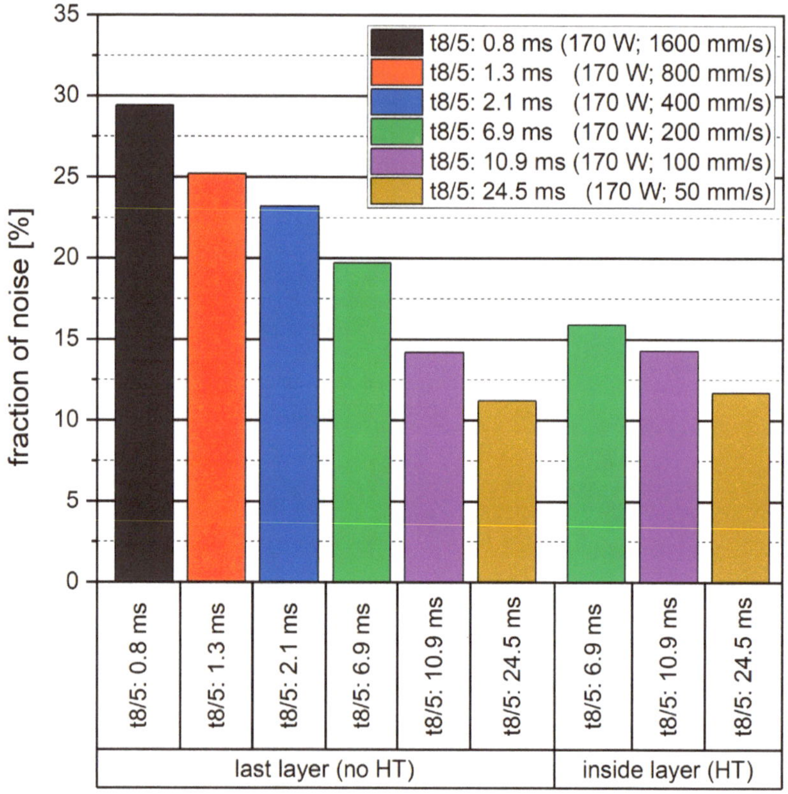

Figure 18. Correlation of cooling duration and fraction of noise within EBSD measurements of heat-treated (HT) and non-heat-treated (no HT) single laser tracks. Value is presenting the ratio of areas that are assumed to present areas of extreme deformations or potential amorphization due to high cooling rates.

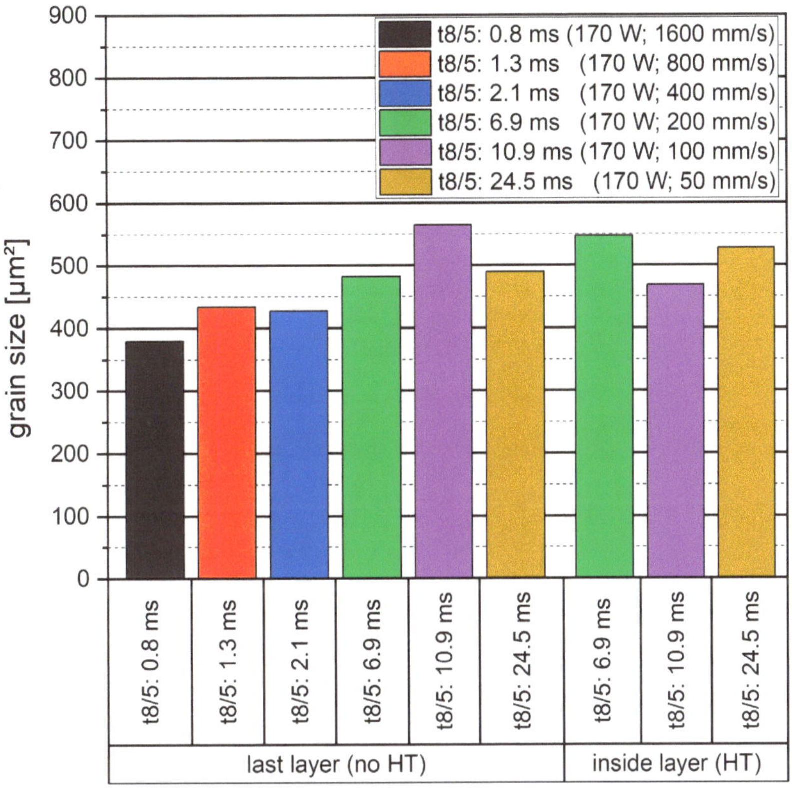

Figure 19. Correlation of cooling duration and grain size measured by EBSD for heat-treated (HT) and non-heat-treated (no HT) single laser tracks.

4. Conclusions

The findings of this work prove that the cooling rate in L-PBF can be measured by pyrometry. A novel approach is introduced in order to overcome the observed problem of signal disturbing by-products within the measurement zone. The four following main topics have been addressed in order to realize the first systematic investigations regarding the correlation of exposure parameters, cooling rates and resulting microstructure:

1. **Quantitative measurements by pyrometry within L-PBF have been realized:** Two different approaches have been introduced in order to correlate the pyrometer signal to a temperature-dependent emissivity within the L-PBF system. The material's emissivity has been measured up to 938 °C (see Figures 5 and 7). These data can serve as a calibration table for future measurements by one-color pyrometry. Furthermore, the use of two-color pyrometry has been introduced, showing a maximum deviation of ±2%.
2. **Introducing a novel approach in order to measure cooling durations of single laser tracks:** Within the single laser tracks experiments, by-products have been identified as the main source of error in measuring the cooling behavior. In fact, measurement within the actual melt pool is not possible. A newly developed and validated fitting method has been introduced, enabling statements about the cooling rates based on incomplete measurement data due to by-products.
3. **Systematic investigation of cooling duration** Based on the approach above, systematic investigations regarding the cooling duration have been carried out. The results show the correlation between cooling duration and process parameters such as laser

power, scanning speed, laser focus size and laser focus offset. The gained quantitative information can be used for simulation validation as well as the understanding of microstructural evolution and can therefore support the future development of FGMs.

4. **Observed phenomena in microstructure** EBSD measurements on single laser tracks reveal wide areas of noise. It is concluded that these areas represent either severe deformation or potential amorphization due to extremely high cooling rates. The fraction of noise within the laser tracks correlates with the cooling duration and therefore supports the hypothesis above.

The results support the validation of simulation results as well as the development and optimization of exposure parameters and resulting mechanical properties. The knowledge gained about the cooling duration supports the understanding of microstructural formation mechanisms. In addition, this work presents the foundation for future two-dimensional pyrometry experiments. These would support a deeper understanding of annealing effects within the L-PBF process, which result from the exposure of the following process layers. The results of this work also promote the use of pyrometry as a tool for process monitoring.

Author Contributions: Conceptualization, A.P.; methodology, A.P.; validation, A.P. and S.S.; formal analysis, A.P.; investigation, A.P., M.J. and S.S.; resources, A.P. and S.S.; data curation, A.P. and M.J.; writing—original draft preparation, A.P.; writing—review and editing, S.S., M.J. and F.B.; visualization, A.P.; supervision, F.B.; project administration, A.P.; funding acquisition, A.P. All authors have read and agreed to the published version of the manuscript.

Funding: This research was funded by Bundeswehr Research Institute for Materials Fuels and Lubricants (WIWeB), Erding, Germany grant number E/E210/AK030/GF126.

Data Availability Statement: Data will be made available upon request.

Conflicts of Interest: The authors declare no conflict of interest. The funders had no role in the design of the study; in the collection, analyses, or interpretation of data; in the writing of the manuscript; or in the decision to publish the results.

References

1. Wohlers, T.T.; Campbell, I.; Diegel, O.; Huff, R.; Kowen, J. *Wohlers Report 2022: 3D Printing and Additive Manufacturing Global State of the Industry*; Wohlers Associates: Fort Collins, CO, USA, 2022; ISBN 978-0-9913332-9-5.
2. McCann, R.; Obeidi, M.A.; Hughes, C.; McCarthy, É.; Egan, D.S.; Vijayaraghavan, R.K.; Joshi, A.M.; Acinas Garzon, V.; Dowling, D.P.; McNally, P.J.; et al. In-situ sensing, process monitoring and machine control in Laser Powder Bed Fusion: A review. *Addit. Manuf.* **2021**, *45*, 102058. [CrossRef]
3. Popovich, V.A.; Borisov, E.V.; Popovich, A.A.; Sufiiarov, V.; Masaylo, D.V.; Alzina, L. Functionally graded Inconel 718 processed by additive manufacturing: Crystallographic texture, anisotropy of microstructure and mechanical properties. *Mater. Des.* **2017**, *114*, 441–449. [CrossRef]
4. Pfaff, A.; Jäcklein, M.; Hoschke, K.; Wickert, M. Designed Materials by Additive Manufacturing—Impact of Exposure Strategies and Parameters on Material Characteristics of AlSi10Mg Processed by Laser Beam Melting. *Metals* **2018**, *8*, 491. [CrossRef]
5. Kürnsteiner, P.; Wilms, M.B.; Weisheit, A.; Gault, B.; Jägle, E.A.; Raabe, D. High-strength Damascus steel by additive manufacturing. *Nature* **2020**, *582*, 515–519. [CrossRef]
6. Everton, S.K.; Hirsch, M.; Stravroulakis, P.; Leach, R.K.; Clare, A.T. Review of in-situ process monitoring and in-situ metrology for metal additive manufacturing. *Mater. Des.* **2016**, *95*, 431–445. [CrossRef]
7. Hou, Z.-J.; Wang, Q.; Zhao, C.-G.; Zheng, J.; Tian, J.-M.; Ge, X.-H.; Liu, Y.-G. Online Monitoring Technology of Metal Powder Bed Fusion Processes: A Review. *Materials* **2022**, *15*, 7598. [CrossRef]
8. Yang, T.; Mazumder, S.; Jin, Y.; Squires, B.; Sofield, M.; Pantawane, M.V.; Dahotre, N.B.; Neogi, A. A Review of Diagnostics Methodologies for Metal Additive Manufacturing Processes and Products. *Materials* **2021**, *14*, 4929. [CrossRef]
9. Mohr, G.; Nowakowski, S.; Altenburg, S.J.; Maierhofer, C.; Hilgenberg, K. Experimental Determination of the Emissivity of Powder Layers and Bulk Material in Laser Powder Bed Fusion Using Infrared Thermography and Thermocouples. *Metals* **2020**, *10*, 1546. [CrossRef]
10. Pavlov, M.; Doubenskaia, M.; Smurov, I. Pyrometric analysis of thermal processes in SLM technology. *Phys. Procedia* **2010**, *5*, 523–531. [CrossRef]
11. Gutknecht, K.; Cloots, M.; Sommerhuber, R.; Wegener, K. Mutual comparison of acoustic, pyrometric and thermographic laser powder bed fusion monitoring. *Mater. Des.* **2021**, *210*, 110036. [CrossRef]
12. Xiao, X.; Liu, X.; Cheng, M.; Song, L. Towards monitoring laser welding process via a coaxial pyrometer. *J. Mater. Process. Technol.* **2020**, *277*, 116409. [CrossRef]

13. Hooper, P.A. Melt pool temperature and cooling rates in laser powder bed fusion. *Addit. Manuf.* **2018**, *22*, 548–559. [CrossRef]
14. Scipioni Bertoli, U.; Guss, G.; Wu, S.; Matthews, M.J.; Schoenung, J.M. In-situ characterization of laser-powder interaction and cooling rates through high-speed imaging of powder bed fusion additive manufacturing. *Mater. Des.* **2017**, *135*, 385–396. [CrossRef]
15. Altenburg, S.J.; Scheuschner, N.; Straße, A.; Gumenyuk, A.; Maierhofer, C. *Towards the Determination of Real Process Temperatures in the LMD Process by Multispectral Thermography*; SPIE: Washington, DC, USA, 2021.
16. Farshidianfar, M.H.; Khajepour, A.; Gerlich, A.P. Effect of real-time cooling rate on microstructure in Laser Additive Manufacturing. *J. Mater. Process. Technol.* **2016**, *231*, 468–478. [CrossRef]
17. Takata, N.; Liu, M.; Li, H.; Suzuki, A.; Kobashi, M. Fast scanning calorimetry study of Al alloy powder for understanding microstructural development in laser powder bed fusion. *Mater. Des.* **2022**, *219*, 110830. [CrossRef]
18. Artzt, K.; Siggel, M.; Kleinert, J.; Riccius, J.; Requena, G.; Haubrich, J. Pyrometric-Based Melt Pool Monitoring Study of CuCr1Zr Processed Using L-PBF. *Materials* **2020**, *13*, 4626. [CrossRef]
19. Lin, Z.; Lai, Y.; Pan, T.; Zhang, W.; Zheng, J.; Ge, X.; Liu, Y. A New Method for Automatic Detection of Defects in Selective Laser Melting Based on Machine Vision. *Materials* **2021**, *14*, 4175. [CrossRef]
20. Grasso, M. In Situ Monitoring of Powder Bed Fusion Homogeneity in Electron Beam Melting. *Materials* **2021**, *14*, 7015. [CrossRef]
21. Thombansen, U.; Gatej, A.; Pereira, M. Process observation in fiber laser–based selective laser melting. *Opt. Eng* **2015**, *54*, 11008. [CrossRef]
22. Yamazaki, K.; Yamamoto, E.; Suzuki, K.; Koshiishi, F.; Tashiro, S.; Tanaka, M.; Nakata, K. Measurement of surface temperature of weld pools by infrared two colour pyrometry. *Sci. Technol. Weld. Join.* **2010**, *15*, 40–47. [CrossRef]
23. Book, A. Principle, Advantages, Limitations and Applications of Two-Colour Pyrometers in Thermal Processes: Technical Report Two-colour Pyrometer. 2018. Available online: https://pdf.directindustry.com/pdf/keller-its-infrared-temperature-solutions/tr-two-colour-pyrometer-201507-en/14914-705255.html (accessed on 24 March 2023).
24. Karayagiz, K.; Elwany, A.; Tapia, G.; Franco, B.; Johnson, L.; Ma, J.; Karaman, I.; Arróyave, R. Numerical and experimental analysis of heat distribution in the laser powder bed fusion of Ti-6Al-4V. *IISE Trans.* **2019**, *51*, 136–152. [CrossRef]
25. Krakhmalev, P.; Yadroitsava, I.; Fredriksson, G.; Yadroitsev, I. In situ heat treatment in selective laser melted martensitic AISI 420 stainless steels. *Mater. Des.* **2015**, *87*, 380–385. [CrossRef]
26. Li, Z.; Li, H.; Yin, J.; Li, Y.; Nie, Z.; Li, X.; You, D.; Guan, K.; Duan, W.; Cao, L.; et al. A Review of Spatter in Laser Powder Bed Fusion Additive Manufacturing: In Situ Detection, Generation, Effects, and Countermeasures. *Micromachines* **2022**, *13*, 1366. [CrossRef] [PubMed]
27. Luo, N.; Scheitler, C.; Ciftci, N.; Galgon, F.; Fu, Z.; Uhlenwinkel, V.; Schmidt, M.; Körner, C. Preparation of Fe-Co-B-Si-Nb bulk metallic glasses by laser powder bed fusion: Microstructure and properties. *Mater. Charact.* **2020**, *162*, 110206. [CrossRef]

materials

MDPI

Article

New Partially Water-Soluble Feedstocks for Additive Manufacturing of Ti6Al4V Parts by Material Extrusion

Ralf Eickhoff [1], Steffen Antusch [1], Dorit Nötzel [1,2] and Thomas Hanemann [1,2,*]

1 Institute for Applied Materials, Karlsruhe Institute of Technology, Hermann-von-Helmholtz-Platz 1, D-76344 Eggenstein-Leopoldshafen, Germany
2 Department of Microsystems Engineering, University Freiburg, Georges-Koehler-Allee 102, D-79110 Freiburg, Germany
* Correspondence: thomas.hanemann@kit.edu or thomas.hanemann@imtek.de; Tel.: +49-721-608-22585

Abstract: In this work, a process chain for the realization of dense Ti6Al4V parts via different material extrusion methods will be introduced applying eco-friendly partially water-soluble binder systems. In continuation of earlier research, polyethylene glycol (PEG) as a low molecular weight binder component was combined either with poly(vinylbutyral) (PVB) or with poly(methylmethacrylat) (PMMA) as a high molecular weight polymer and investigated with respect to their usability in FFF and FFD. The additional investigation of different surfactants' impact on the rheological behaviour applying shear and oscillation rheology allowed for a final solid Ti6Al4V content of 60 vol%, which is sufficient to achieve after printing, debinding and thermal densification parts with densities better than 99% of the theoretical value. The requirements for usage in medical applications according to ASTM F2885-17 can be fulfilled depending on the processing conditions.

Keywords: material extrusion; FFF; FDM; FFD; additive manufacturing; titanium alloys; Ti6Al4V; water-soluble binder; feedstock

check for updates

Citation: Eickhoff, R.; Antusch, S.; Nötzel, D.; Hanemann, T. New Partially Water-Soluble Feedstocks for Additive Manufacturing of Ti6Al4V Parts by Material Extrusion. *Materials* **2023**, *16*, 3162. https://doi.org/10.3390/ma16083162

Academic Editors: Swarup Bag and Christ Prakash Paul

Received: 1 March 2023
Revised: 2 April 2023
Accepted: 13 April 2023
Published: 17 April 2023

1. Introduction

Additive manufacturing, quite often also denoted as 3D printing, is nowadays an established fabrication method of polymer, metal and ceramic parts. In most cases, 3D printing is used for rapid prototyping, but these methods have gained more and more importance for rapid tooling and even rapid manufacturing of small-scale series [1–8]. Due to their history of invention and the operation principle, the most important techniques are VAT photopolymerization such as stereolithography (SLA), material extrusion (MEX) such as fused filament fabrication (FFF), and powder bed fusion such as selective laser melting (SLM) or electron beam melting (EBM). In recent years, a pronounced material and process development widened the initial material limitations and enabled the adaptation of SLA and MEX-based printing techniques for the processing of ceramic and metal materials. FFF and related techniques benefit from the long feedstock development experience in powder injection molding (PIM) for the fabrication of dense and warpage-free metal and ceramic parts according to the following process chain [9]:

1. Feedstock compounding of using suitable binders, additives and a huge solid load (ceramics: ≥50 vol%; metals: ≥60 vol%) considering spherical fillers ideally;
2. Printing via MEX techniques;
3. Debinding, often a combination of liquid pre-debinding with thermal debinding;
4. Sintering with conditions depending on the filler's properties.

With respect to the knowledge gained from PIM, several groups were able to realize different ceramics and metals applying MEX methods [9–19]. Perhaps due to the more complex sintering step, only a few reports dealing with FFF of metals were published [20–26]. Currently, lightweight and medical applications are upcoming fields of interest, especially using titanium alloys such as Ti6Al4V. The application of the latter as an implant material is outlined in the ASTM F2885-17 standard, describing, amongst others, the relevant mechanical properties [27]. The additive manufacturing applying Ti6Al4V via powder-based methods such as SLS, SLM or EBM has been reported and is commercialized partially [28–31]; quite recently, MEX methods were also used for part realization [32–36].

A comprehensive investigation of the feedstock flow behaviour is essential to meet the two contradictory requirements of a suitable feedstock. On the one hand, the solid load must be as high as possible to minimize shrinkage and warpage risk during sintering. On the other hand, the solid load must not be too high to allow a good melt flow behaviour, enabling a defect-free extrusion and filament deposition without additional voids due to solid overfilling. A highly filled feedstock with a solid particle loading represents a complex system with the occurrence of elastic (by the solid) and viscous (by the polymer) properties in the molten state. With respect to a proper description of the melt flow behaviour, it is necessary to measure in addition to the shear rate dependent viscosity the viscoelastic flow via oscillation measurements applying a rotary rheometer. The latter technique was exercised for different feedstock systems [17,18,37], and quite recently in [36]. It is possible to determine the complex shear modulus G* containing the contributions of the storage modulus G′ (elastic part) and the loss modulus G″ (viscous part) (1). With the knowledge of the applied angular frequency, the complex viscosity can be estimated (2). The direct relationship between the complex shear modulus and the complex viscosity is shown in (3) [38]. An accurate determination of the viscoelastic flow and the feedstocks yield point is of particular importance for feedstock and printing parameter development with respect to proper and defect-free deposition of a freshly printed layer on the previous one. A more detailed description of the relevance of the complex shear modulus G″ can be found in [36] and especially in [38].

$$G^* = G' + iG'' \tag{1}$$

$$\eta^* = \eta' + i\eta'' = \frac{G'}{\omega} + i\left(-\frac{G''}{\omega}\right) \tag{2}$$

$$|\eta^*| = \left|\frac{G^*}{\omega}\right| \tag{3}$$

The following work extends the research on suitable feedstock systems for Ti6Al4V from the initially used nonpolar wax/polyethylene (PE) mixtures to polar and therefore partially water-soluble binder systems, which simplifies especially the solvent-based pre-debinding towards eco-friendly solvents such as water, instead of polluting hexane/heptane solvents [39].

2. Materials and Methods

2.1. Polar Feedstock Development

With respect to potential applications in health care products such as implants and in continuation of previous published work [36], Ti6Al4V (Grade 23FE, Heraeus, Hanau, Germany) was selected as a metal filler [40]. Due to the demanding mechanical property requirements in medical engineering, this grade shows a low carbon and oxygen content according to the ASTM F2885-17 standard [27], derived from the specifications for metal injection molding (MIM) produced Ti6Al4V surgical implants.

In significant contrast to previously published work [36], two polar binder systems consisting of the low molecular weight polyethylene glycol (PEG) and the high molecular weight polymers poly(vinylbutyral) (PVB) or poly(methylmethacrylate) (PMMA) have been investigated (Figure 1). Due to synthesis conditions, PVB can be described as a copolymer containing the aspired butyral group, an acetyl- as well as the highly polar hydroxy functionality. Both binders, PEG/PVB and PEG/PMMA, have been successfully applied for ceramics in FFF and injection molding, enabling crack-free and dense alumina or zirconia parts [12,13,41–44]. In this work, PEG with different average molar weights M_W (2000, 4000, 6000, 8000, 20,000 g/mol) (C. Roth GmbH Co KG, Karlsruhe, Germany) have been applied. PVB Mowital B30H (Kuraray Europe GmbH, Frankfurt, Germany) and PMMA Degalan G7E (Roehm GmbH, Darmstadt, Germany) were selected as high molecular weight polymers. Stearic acid (SA, C. Roth GmbH Co KG, Karlsruhe, Germany) as well as two commercial additives, PAT-77/P and PAT-659/CB (E. & P. Wuertz GmbH Co. KG, Bingen, Germany), with proprietary composition, recommended by the vendor for polymer/metal feedstocks, have been used as surfactants. Normally, the surfactant amount is directly related to the specific surface area of the filler, but due to the low specific surface area (SSA) of the used Ti6Al4V, a fixed surfactant concentration of 5 wt% of the whole binder amount is applied and subtracted from the low molecular weight PEG. The thermal behaviour of all investigated organic binder components can be found in Table 1.

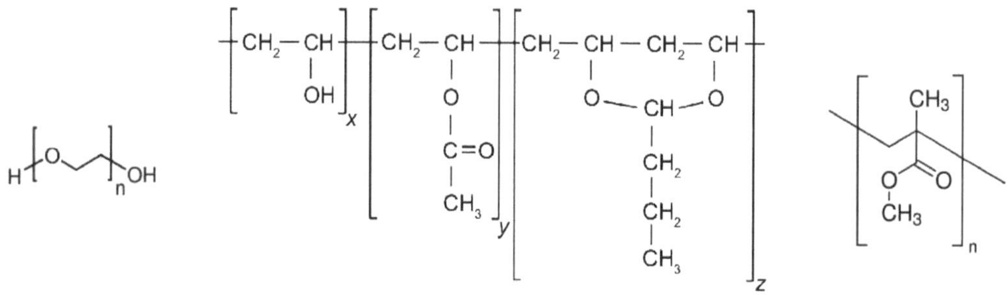

(a) (b) (c)

Figure 1. Chemical structures of the used main binder components: (**a**) polyethylene glycol (PEG); (**b**) poly(vinylbutyral) (PVB); (**c**) poly(methylmethacrylate) (PMMA).

Table 1. Overview of the used binder components (T_S: softening temperature, T_D: decomposition temperature).

Component	Density (g/cm^3)	M_W (g/mol)	T_S (°C)	T_D (°C)
PEG 2000	1.22 ^{M}	1800–2200 ^{V}	52 ^{V}	n.a.
PEG 4000	1.22 ^{M}	3500–4500 ^{V}	58 ^{V}	n.a.
PEG 6000	1.22 ^{M}	5000–7000 ^{V}	60 ^{V}	n.a.
PEG 8000	1.22 ^{M}	7300–9000 ^{V}	60 ^{V}	n.a.
PEG 20,000	1.22 ^{M}	20,000 ^{V}	60 ^{V}	n.a.
PVB 30 H	1.11 ^{M}	32,000–35,000 ^{V}	68 ^{V}	206 ^{M}
PMMA Degalan 7 E	1.18 ^{M}	159,000 ^{M}	96–109 ^{M}	240 ^{M}
Stearic acid	0.92 ^{M}	284 ^{V}	69–85 ^{V}	170 ^{M}
PAT-77 P	0.97 ^{M}	n.a.	85–95 ^{V}	n.a.
PAT-659/CB	0.96 ^{M}	n.a	45–75 ^{V}	n.a.

^{M}: measured data, ^{V}: vendor's data sheet.

In addition to the data submitted by the vendors, a comprehensive educt characterization is mandatory, especially in the case of the selected Ti6 Al4V filler. With respect to feedstock development, the following four main particle properties must be characterized:

- Particle size distribution;
- Specific surface area;
- Particle density;
- Particle morphology.

The particle size distribution was measured via laser diffraction (LA-950 Horiba Ltd., Kyoto, Japan), the specific surface area (SSA) by the Brunauer-Emmett-Teller adsorption method (BET, Gemini VII 2390, Micromeritics Instr. Corp., Norcross, GA, USA), the particle density by helium pycnometry (Pycnomatic ATC, Porotec, Germany) and, finally, the particle morphology with a SEM Supra 55 (Zeiss, Oberkochen, Germany).

2.2. Feedstock Preparation

A torque recording mixer-kneader (W50-EHT, Brabender GmbH, Duisburg, Germany) was applied for the new polar feedstock compounding. The blade's rotating speed was set to 30 rpm, which is equivalent to a maximum shear rate of 36 1/s for 1 h compounding time. The compounding temperature was set to 110 °C (PEG/PVB) and 160–170 °C (PEG/PMMA) according to the different glass transition temperatures of the high molecular weight polymer. Further details can be found in [36].

2.3. Rheological Characterization

All feedstock systems were investigated by two common established rheometric methods. The temperature and shear rate dependent viscosity were measured by a high-pressure capillary rheometer (Rheograph 25, Goettfert Werkstoff-Pruefmaschinen GmbH, Buchen, Germany). For better comparison with other feedstock systems [36], identical measuring parameters (temperature: 160 °C; capillary diameter and length: 1 and 30 mm, shear rate range: 1 to 5000 1/s) were used. With respect to the determination of the complex flow behaviour, especially the detection of potential yield points as a function of the feedstock composition, oscillatory measurements in the amplitude sweep (AS) mode applying a rotary rheometer (Gemini HR Nano, Netzsch-Geraetebau GmbH & Co., Selb, Germany) equipped with a plate-plate geometry (20 mm diameter, gap 1 mm, smooth surface) were applied. To be consistent with previously reported work, identical measuring parameters (temperatures, frequency, shear stress range) were used [36].

2.4. Material Extrusion (MEX)

2.4.1. Fused Filament Fabrication (FFF)

After feedstock compounding and pelletizing with an impact mill (Granulator 1514, Rapid Germany, Kleinostheim, Germany), the granules were reshaped into filaments by a single screw filament extruder (Noztek pro HT, Noztek, Shoreham, UK). All FFF-printing trials were performed with a German RepRap X350pro printer (Feldkirchen, Germany) applying different extruder nozzles (0.15, 0.3 and 0.4 mm). With respect to acceptable printing results, some machine modifications were necessary; details can be found in [36].

2.4.2. Fused Feedstock Deposition (FFD)

In the case of FFD printers, pellets or granules can be used instead of filaments. The two-component FFD printer FFD 150H (3d-figo GmbH, Salzkotten, Germany) carries two small extruders placed directly above the printheads. In all cases, a nozzle diameter of 0.4 mm was used. For better material adhesion and printed part removal, the printing platform was covered with PE-coated spring steel. It was necessary to preheat the feedstock pellets in the extruder to accelerate the extrusion process. Two different slicers for machine control were applied: Ultimaker Cura and Repetier-Host (V 2.2.2).

2.5. Debinding, Sintering and Hot Isostatic Pressing

The debinding step was subdivided into two parts. Initially, the PEG was removed in water (40 °C, 24 h), which generates open pores. For this, the printed parts were placed on a die plate in a glass reactor setup. Then, thermal debinding was conducted prior to sintering in a metal sinter furnace (MUT Advanced Heating GmbH, Jena, Germany) with a maximum debinding temperature of 550 °C, subsequently followed by sintering (max. temperature 1350 °C). The thermal treatment was performed under argon atmosphere to avoid further oxygen contamination. In the case of selected samples and with respect to enhanced density values, hot isostatic pressing (HIP, HIP3000, Dieffenbacher GmbH, Eppingen, Germany) was applied.

2.6. Characterization of Sintered Samples

A set of different techniques were used for a comprehensive sample characterization covering density measurements following the Archimedes principle (Secura 225D-1S equipped with YDK 01, Sartorius Lab Instruments GmbH & Co KG, Göttingen, Germany) and metallographic preparation described in detail in [36]. With respect to the final carbon and oxygen content in the sintered part, a combustion analysis (CS600 and TC600, LECO Instruments GmbH, Mönchengladbach, Germany) was performed.

3. Results and Discussion

3.1. Material Characterization

From previous work dealing with powder injection molding, FFF or FFD in the case of ceramic or metal part fabrication, a huge solid load is necessary to achieve dense parts (ceramic: >99% theoretical density, metals: >98% theoretical density). The presence of spherical particles is mandatory to obtain moderate viscosity values. The morphology of the applied Ti6Al4V powder is spherical with a monomodal distribution around 30 μm, a density around 4.4 g/cm^3 and a SSA value of 0.15 m^2/g [36], which is quite small in contrast to established ceramic materials such as alumina or zirconia.

3.2. Feedstock Compounding

3.2.1. PEG/PVB Binder System

PEG/PVB binder systems in combination with alumina or zirconia fillers and SA as a surfactant were previously used in ceramic injection molding [41–43] and in FFF [12,13]. Here, the influence of three different surfactants on compounding has been investigated. Figure 2a shows the time-depending compounding torque of the PEG/PVB mixtures without and with the three different investigated surfactants applying a PEG M_W of 8000 g/mol. In general, the compounding curve can be split in three phases, namely the filling phase, the mixing phase and, finally, the equilibrium phase [45]. From the torque vs. time diagram the quality of the resulted feedstock can be evaluated, e.g., a constant final equilibrium torque value is a strong hint for a homogenous feedstock composition [45]. The observed compounding behaviour is slightly different from the previously investigated wax/PE mixtures [36]. The applied solid load of 60 vol% yielded a low maximum torque value of around 12 Nm (kneaded at 110 °C), which is almost twice the value of the related wax/PE mixtures (kneaded at 125 °C) [36]. After passing the mixing stage, a pronounced torque drop can be seen followed by a slight increase in all cases; only the usage of SA delivers an almost constant torque value. The compounding behaviour was almost identical with and without a surfactant. The addition of the two PAT surfactants caused a slight torque increase. The negligible surfactant influence can be attributed to the very small SSA and the spherical particle shape. In the literature, some other binder systems were employed for Ti6Al4V as a solid filler:

- Unknown proprietary binder with solid load of 59 vol% [33,34];
- Polyethylene vinyl acetate/poly(propylene ethylene)/polyisobutene with SA as a surfactant [35], solid load 60–65 vol%;
- Poly(propylene ethylene)/maleic acid anhydride functionalized polypropylene with a solid load up to 60 vol% [37].

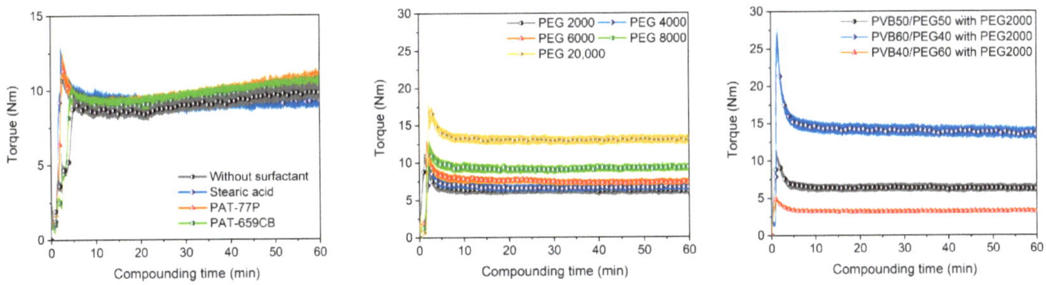

Figure 2. Compounding of PEG/PVB-based binder (110 °C, 60 vol%) with different compositions. (**a**) PEG 8000 as low molecular component and with or without different surfactants (5 wt%); (**b**) variation of the PEG M_W with 5 wt% SA as surfactant; (**c**) variation of PEG/PVB ratio applying PEG 2000 as low molecular binder part and 5 wt% SA as surfactant.

The average molecular weight of the used PEGs had a clear impact on the compounding behaviour. While the equilibrium torque values of the low molecular weight PEG (2000, 4000, 6000, 8000 g/mol) were close together, a further increase to 20,000 g/mol caused almost a torque doubling (Figure 2b). This can be attributed to the increasing probability of chain entanglement accompanied by increasing inner friction. The variation of the PEG/PVB ratio can also be used for viscosity adjustment. Applying PEG 2000 as low molecular weight binder part, the increase of the PEG 2000 fraction yielded a pronounced torque drop and a faster access to the equilibrium state (Figure 2c).

3.2.2. PEG/PMMA Binder System

PEG/PMMA systems have been also introduced as a binder in PIM and material extrusion. The influence of three PMMAs with different M_W and granule morphology on compounding, melt rheology and micro powder injection molding of zirconia and the final ceramic density was described in [44,46]. A binder, consisting of PEG (M_W: 1500 g/mol) and a PMMA emulsion with SA as a surfactant, was used in metal injection molding of tungsten carbide/cobalt mixtures [47]. The liquid debinding behaviour of PEG/PMMA feedstocks in water with Ti6Al4V as a filler and different SA contents is described in [48]. The binder possesses a huge excess of PEG (M_W: 4000 g/mol) relative to PMMA of almost 9:1, simplifying the dissolution process [48]. Hayat et al. investigated the influence of the PEG M_W (1500–20,000 g/mol) on the flow and water debinding behaviour of titanium-filled PEG/PMMA feedstocks for metal injection molding [49]. The higher the average molecular weight, a higher viscosity and slower solvent debinding in water can be observed [49]. As in the previous case, the influence of the three different surfactants (5 wt%) on the compounding process were investigated; again, the surfactant replaced partially the PEG moiety. Figure 3a presents the torque as function of time and used surfactant (PEG 8000). Due to the huge glass transition temperature of the PMMA (96–109 °C), a compounding temperature of 170 °C was necessary. The compounding curve (Figure 3a) looks quite different than in the PEG/PVB system shown in Figure 2a and can be attributed to the shape of the PMMA pellets and the higher glass transition temperature. Prior to mixing, the pellets have to be liquified, which can be seen in the pronounced extension of the filling and mixing phase up to 30 min instead of 5 min in the PEG/PVB binder. The influence of the surfactants is more obvious; the PAT-659/CB especially caused a significant initial torque increase. After 1 hour compounding time, only the addition of SA delivered a constant torque value. The M_W of the used PEGs had only a small impact on the compounding behaviour (Figure 3b). As expected, an increasing M_W caused a slight torque increase; the compounding process is more dominated by the pellet melting.

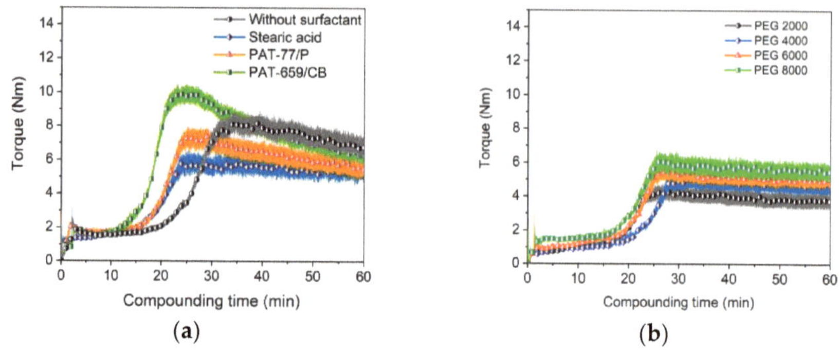

Figure 3. Compounding of PEG/PMMA-based binder (170 °C, 60 vol%). (**a**) PEG 8000 as low molecular component and with or without different surfactants (5 wt%) at a 1 to 1 PEG/PMMA ratio; (**b**) variation of the PEG M_W (5 wt% SA).

Following Table 1, a compounding temperature of 170 °C is quite high and, e.g., very close to the decomposition temperature of SA. Therefore, a feedstock processing at lower temperatures is recommended, which can be achieved by viscosity reduction due to an increase of the binder amount with lower M_W [49]. Figure 4a–c show for the different surfactants the change of the torque with proceeding compounding time and variable PMMA to PEG (here PEG 8000) ratio. In all cases, a pronounced maximum torque drop can be observed, the impact of the PMMA liquification shrinks with the reduction in the PMMA amount and the low viscosity of the PEG 8000 dominates. The differences between the surfactants diminishes as well with higher PEG concentrations. Therefore, the PEG amount increase allows for a compounding temperature reduction or for higher solid loadings. Both effects—the temperature drop and solid load increase—are depicted in Figure 5. Figure 5a shows the influence of the temperature on compounding applying a feedstock with a PMMA/PEG8000 ratio of 40/60 and a solid load of 60 vol% Ti6Al4V. A higher compounding temperature yields in a lower torque as well as in a faster equilibrium stage. At a higher load of 65 vol% (Figure 5b), a temperature increase helps in the same manner, but under consideration of an enhanced SA decomposition.

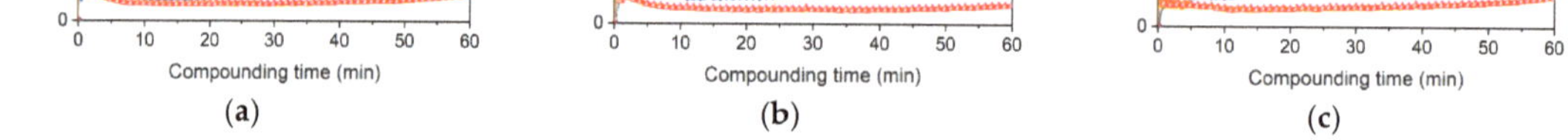

Figure 4. Compounding using different surfactants (5 wt%). (**a**) SA; (**b**) PAT-77P; (**c**) PAT-659/CB.

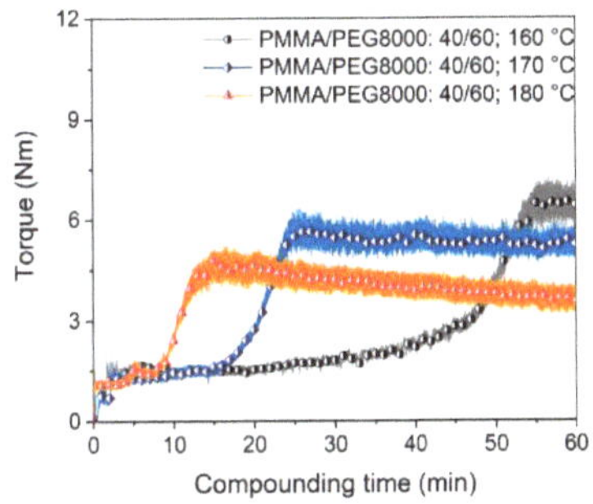

(a)

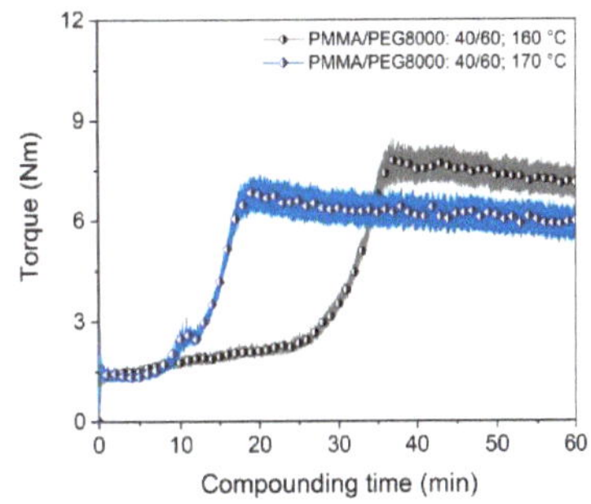

(b)

Figure 5. Compounding of PEG/PMMA-based binder and SA as surfactant. (**a**) PMMA/PEG 8000 ratio 40/60 at different temperatures (solid load 60 vol%); (**b**) PMMA/PEG 8000 ratio 40/60 at different temperatures (solid load 65 vol%).

3.3. Rheological Characterization

3.3.1. PEG/PVB Binder with Constant PEG/PVB Ratio

As described previously [36], a comprehensive description of the feedstocks' melt flow behaviour is crucial for the successful usage in FFF, FFD as well as in injection molding. Figure 6a shows for the mixture PEG/PVB and different surfactants the melt flow behaviour at a solid load of 60 vol% and a temperature of 160 °C. A viscosity measurement without surfactant was not possible due to phase separation at moderate and higher shear rates. Whilst the presence of SA and PAT-77/P caused the typical pseudoplastic (shear-thinning) flow with almost identical viscosity values, the addition of PAT659/CB resulted in an unusual melt flow with a viscosity increase at larger shear rates representing slight dilatancy. This phenomenon was reproducible. Dilatancy normally occurs if a superstructure with increasing attractive forces is formed and should be attributed here to the molecular structure of the surfactant, which is unfortunately proprietary. The viscosities were comparable to the ones applying wax/LDPE feedstocks at the same solid load and identical measuring temperature [36]. The variation of the PEG M_W had an impact on the resulting viscosity in that increasing M_W yielded a viscosity increase retaining the pronounced pseudoplastic flow, which is favourable for injection molding (Figure 6b). The usage of PEGs with a M_W of at least 6000–8000 g/mol is due to the higher softening temperature advantageous for FFF, FFD and injection molding enabling a higher greenbody stability.

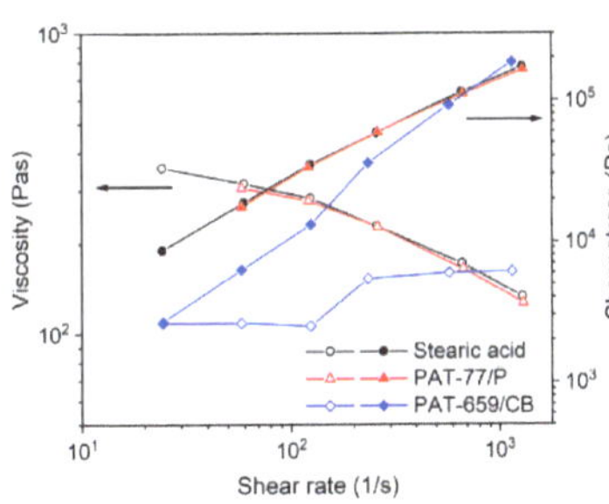

(a)

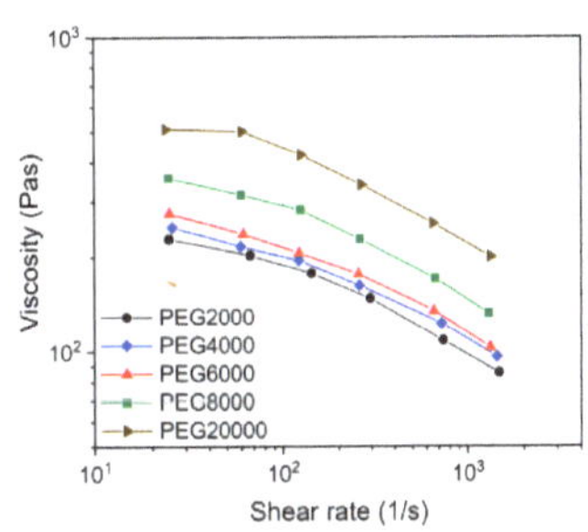

(b)

Figure 6. (**a**) Shear rate dependent melt viscosity as well as shear stress at 160 °C of PEG8000/PVB-based binder at a solid load of 60 vol% and different surfactants; (**b**) variation of the PEGs' average molecular weight with 5 wt% SA as surfactant.

A molten feedstock represents a complex mixture of polar and nonpolar materials and the simultaneous presence of solid and liquid matter causing a viscoelastic flow behaviour, which can be investigated by oscillation rheology measuring the storage G′ and the loss modulus G″ separately, representing the solid-state elastic behaviour (G′) as well as the liquid flow (G″) (Figure 7). The investigated feedstocks containing different surfactants show at low shear stresses a larger storage than loss modulus describing the dominance of the more elastic part of the complex modulus. At a certain shear stress, the two curves cross and the loss modulus dominates, representing the viscous part. This crossover is denominated as the yield or flow point, which is of particular importance for FFF or FFD printing. SA (Figure 7a) shows the smallest complex moduli values of all investigated feedstocks; PAT-77P is at low shear stresses close to SA (Figure 7b), and PAT-659/CB (Figure 7c) possesses the highest complex moduli values. For better comparison, Figure 8a compares the yield point values at 160 °C for the investigated PEG8000/PVB feedstocks with different surfactants. In the case of SA as additive, Figure 8b demonstrates the influence of the PEG M_W on the yield point. Increasing M_W causes a pronounced yield point value raise due to the large possibility of chain entanglement causing enlarged inner friction. These results were supported by a temperature sweep investigation, where a PEG Mw increase from 2000 to 20,000 g/mol yielded a strong complex viscosity increase especially at higher temperatures by more than one decade (Figure 8c).

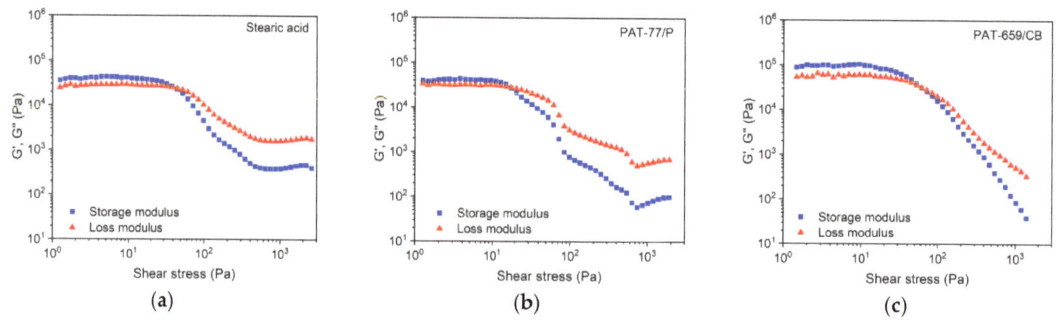

Figure 7. Storage (G′) and loss modulus (G″) of PEG 8000/PVB systems containing three different surfactants (temperature 160 °C, solid load 60 vol%) via oscillatory test applying amplitude sweep. (**a**) SA; (**b**) PAT-77/P; (**c**) PAT-659/CB.

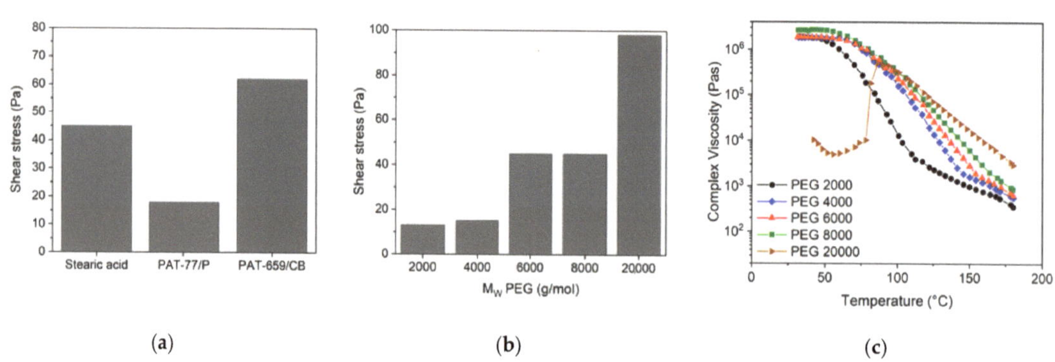

Figure 8. Yield points for PEG/PVB feedstocks (160 °C, solid load 60 vol%). (**a**) Applying PEG 8000 and different surfactants; (**b**) applying PEGs with different M_W and SA as surfactant; (**c**) complex viscosity as function of temperature and PEG with different M_W and SA as surfactant.

3.3.2. PEG/PVB Binder with Different PEG/PVB Ratios

Following the results obtained from compounding, it can be expected that the variation of the PEG/PVB ratio has a pronounced impact on the melt rheology. Figure 9a describes the change of the melt flow as a function of shear rate and PEG2000/PVB ratio. Decreasing PVB amounts allowed a clear viscosity reduction retaining the pseudoplastic flow. The yield point is also affected by the ratio of the low and high molecular binder moiety. Whilst the decrease of a 50%:50% ratio of PEG 2000 to PVB down to 40% PVB has only a small impact on the yield point, the increase up to 60% PVB results in a significant yield point gain by almost a factor of 2.5 (Figure 9b). The variation of the PEG 2000/PVB ratio has an influence on the complex viscosity change with increasing temperature, especially at higher temperatures passing the softening temperatures of PEG 2000 (52 °C) and PVB (68 °C) (Figure 9c).

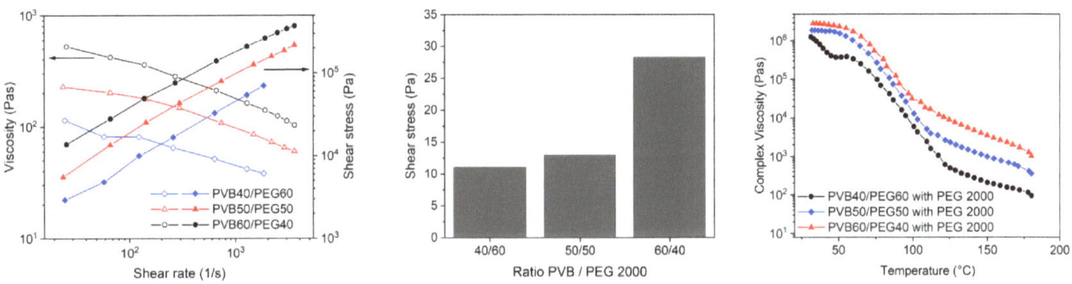

(a) (b) (c)

Figure 9. (**a**) Melt viscosity of PEG 2000/PVB feedstocks with different PEG/PVB ratios (160 °C, solid load 60 vol%); (**b**) yield point dependence on PEG2000/PVB ratio; (**c**) complex viscosity as function of temperature and PEG 2000/PVB ratio.

3.3.3. PEG/PMMA Binder with Constant PEG/PMMA Ratio

In contrast to the related PEG/PVB feedstocks at the same solid load, the melt viscosity is almost higher by one decade, which can be attributed mainly to the huge average molecular weight of the used PMMA (M_W PMMA~159,000 g/mol; PVB~35,000 g/mol, Table 1). The presence of any of the surfactants delivered a small viscosity drop relative to the feedstock without surfactant; the variation of the surfactant type itself did not show any relevant melt flow discrepancy (Figure 10a). The variation of the PEG M_W did not alter the resulting viscosity; remarkably, all systems possessed a typical pseudoplastic flow (Figure 10b).

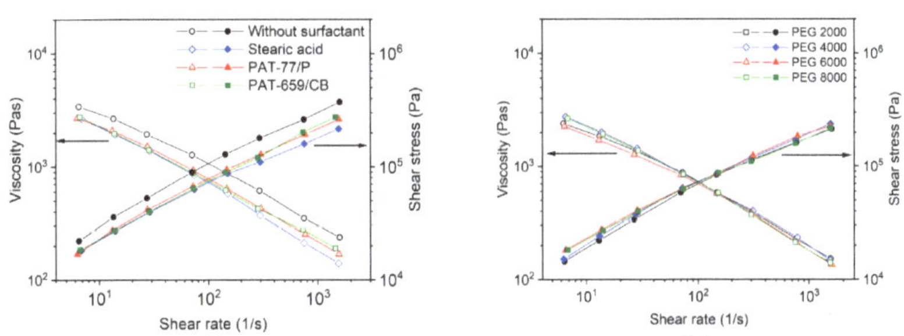

(a) (b)

Figure 10. (**a**) Shear rate dependent melt viscosity as well as shear stress at 160 °C of PEG8000/PMMA-based binder (solid load 60 vol%) and different surfactants; (**b**) variation of the PEG M_W with 5 wt% SA as surfactant (solid load 60 vol%, 160 °C).

As in the PEG/PVB feedstocks, a yield point could be detected by the measurement of the complex viscosity depending on the used surfactant and PEG M_W (Figure 11a,b). All applied surfactants lowered the shear stress value for the yield point in contrast to the feedstock without surfactant (Figure 11a); the two PATs especially had a pronounced impact. The substitution of the PEG 8000 by PEGs with smaller M_W caused a reduction of the yield point shear stress value (Figure 11b). The influence of M_W on the complex viscosity is negligible as well (Figure 11c). These results are in strong contrast to the above presented PEG/PVB feedstocks (Figure 8) and should be attributed to the huge PMMA M_W.

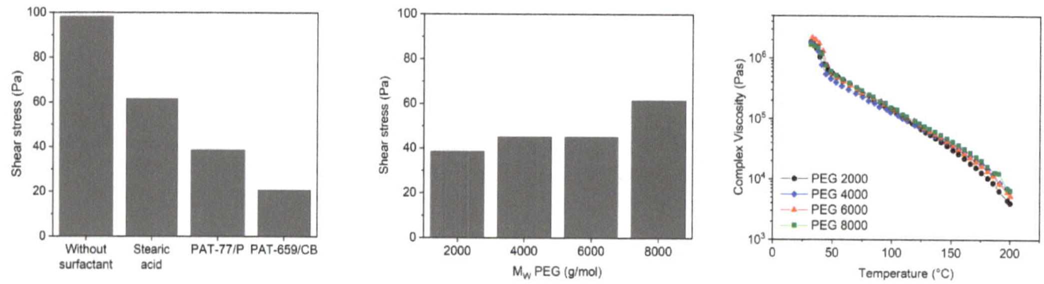

(**a**) (**b**) (**c**)

Figure 11. Shear stress values of the yield points for PEG/PMMA feedstocks (160 °C, solid load 60 vol%). (**a**) Applying PEG 8000 and different surfactants; (**b**) applying PEGs with different M_W and SA as surfactant; (**c**) complex viscosity as function of temperature and PEG with different M_W and SA as surfactant.

3.3.4. PEG/PMMA Binder with Different PEG/PMMA Ratios

In addition to the previously described PEG M_W variation, the flow behaviour can be adjusted by the change of the PEG/PMMA ratio. Increasing PEG 8000 amounts caused a remarkable viscosity drop at constant temperature, especially at concentrations beyond 60% (Figure 12a). A temperature gain from 160 °C up to 200 °C for a system PMMA40/PEG60 (PEG 8000) also enabled a viscosity drop (Figure 12b). The PEG/PMMA ratio variation towards larger PMMA amounts resulted in a yield point shear stress value increase (Figure 12c).

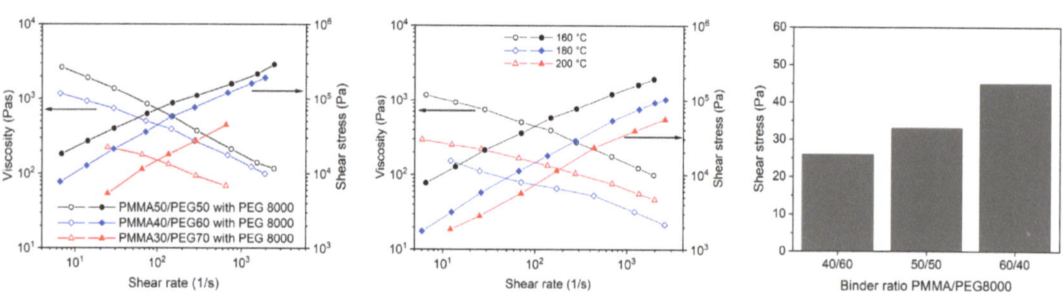

(**a**) (**b**) (**c**)

Figure 12. (**a**) Change of melt flow with increasing PEG 8000 moiety in the binder (solid load; 60 vol%, T = 160 °C, SA as surfactant); (**b**) temperature influence on a PEG60/PMMA40 feedstock applying PEG 8000 (solid load 60 vol%, SA as surfactant); (**c**) variation of the PEG 8000/PMMA ratio and resulting shear stress change for the yield point (SA as surfactant).

3.3.5. Solid Load Variation

Within the PEG/PVB feedstock composition, it was not possible to realize systems with solid loads beyond 60 vol% in an acceptable quality suitable for FFF or FFD printing. With the experience previously gained by the variation of used PEG, the PEG/PMMA ratio, selected surfactant and exploiting the temperature dependence of the melt viscosity, it was possible to realize feedstocks with an enhanced solid load of 65 vol% by an increase of the PEG 8000 moiety. Figure 13 shows the shear stress values for the yield point applying the two PAT surfactants at different temperatures and a solid load of 65 vol%. The values at 160 °C are significantly higher (almost a factor of 10) than the related ones at a solid load of 60 vol%. In previously described results, e.g., in Figure 11, PAT-77P possessed a higher yield point shear stress value than the PAT-659/CB surfactant; the unexpected behaviour at 180 °C may be attributed to phase separation. It has to be noted that the solid load increase from 60 to 65 vol% hampered the feedstock preparation and measurement significantly.

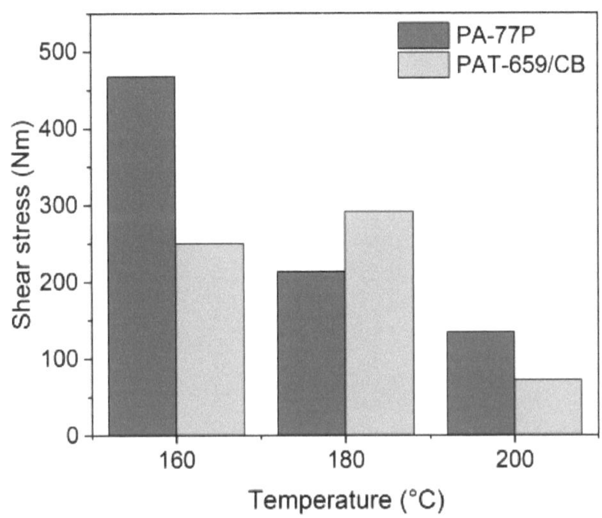

Figure 13. Shear stress values for the yield points for PEG60/PMMA40 (PEG 8000) applying the surfactants PAT-77P and PAT-659/CB at different temperatures (solid load 65 vol%).

3.4. *Material Extrusion (MEX)*

3.4.1. Processing of PEG/PVB Feedstocks

Following the results from rheological characterization, three different PEG/PVB systems with a Ti6Al4V load of 60 vol% and varying M_W of PEG were selected for filament fabrication applying the Noztek filament extruder, equipped with a 2.8 mm nozzle; Table 2 summarizes the achieved filament parameters. All extruded filaments did not show any extrudate swelling directly behind the nozzle. Therefore, the filament diameters are close to the required 2.85–3 mm filament diameter requested by the modified X350Pro FFF printer. This molten filament behaviour is in significant contrast to the previously used wax/LDPE feedstocks [36]. The surface appearance of the filaments was not unique: whilst the PEG 2000 containing filament showed a smooth surface, the two others possessed a rough surface, which may be attributed to phase separation between PEG and PVB and a deteriorated particle wetting.

Table 2. Extruded PEG/PVB filaments.

Feedstock	Extrusion Temperature (°C)	Filament Diameter (mm)
PEG 2000/PVB	90	2.82 ± 0.02
PEG 8000/PVB	90	2.80 ± 0.01
PEG 20,000/PVB	90	2.86 ± 0.01

As in previous work, the filament extrusion delivered rigid filaments which cannot be winded; hence, about 50 cm long straight filaments were extruded on a metal cooling track [36]. These new filaments were then stored in a vacuum oven at 47 °C not less than 24 h to avoid any water adsorption, which is typical for feedstocks with the capability of hydrogen bridge formation, especially PVB [50]. After drying, the filament rods were introduced into the printer's filament extruder from above. The FFF process itself was hampered by the pronounced phase separation at the used printing temperature around 200 °C, especially at low PEG average molecular masses. Table 3 lists the best FFF printing parameters for the different PEG/PVB mixtures applying in all cases SA as a surfactant. Only in the case of the PEG 20,000 the phase separation at low temperatures could be suppressed and devices printed (Figure 14). With the exception of the printing temperature, the printing parameters are almost identical to the previously reported wax/PE feedstocks [36]. Due to the huge viscosity under FFF printing conditions, it was not possible to print parts applying a feedstock with 65 vol%.

Table 3. FFF printing parameters for different PEG/PVB systems (solid load: 60 vol%, 5 wt% SA as surfactant).

FFF Printing Parameter	PEG 2000/PVB	PEG 8000/PVB	PEG 20,000/PVB
Temperature (°C)	230	230	180
Platform temperature (°C)	55	55	60
Printing speed (mm/s)	5	5	5
Printing speed first layer (mm/s)	5	5	2.5
Nozzle diameter (mm)	0.3	0.3	0.3
Infill (%)	105	105	105
Overlap contour lines with infill (mm)	0.2	0.2	0.2

(a) (b)

Figure 14. FFF-printed PEG 20,000/PVB parts. (**a**) Dental implant; (**b**) planetary gear train.

Due to the pronounced phase separation, printing via FFD was impossible according to the higher shear forces at elevated temperatures.

3.4.2. Processing of PEG/PMMA Feedstocks

Because of the comprehensive rheological investigation on the PEG/PMMA feedstocks, the system PEG60/PMMA40 with PEG 8000 as low molecular binder component was selected for FFF printing trials after filament extrusion. Table 4 summarizes the processing parameters; unfortunately, in the case of the PAT659/CB-containing mixture, a straight feedstock track could not be prepared. As in the PEG/PVB systems, an extrudate swelling after passing the nozzle was not observed. The feedstocks with the two different PAT surfactants were suitable for FFF (PAT77-P) and FFD (PAT77-P, PAT659/CB) printing. Surprisingly, SA could not be used due to the phase separation at the printing temperature, which is almost like PEG/PVB feedstocks. It seems to be that at higher temperatures the solubility of SA in PEG/PMMA feedstocks is reduced. Table 5 lists the best printing parameters. In the case of FFD, the printing as well as the platform temperatures are slightly increased in comparison to the ones in FFF, which is attributed to a better adhesion of the first layer on the printing platform. Directly after printing, the parts showed a low greenbody stability, which is caused by the platform temperature close to the softening temperature of the used binder component PEG 8000 (Table 1). At lower temperatures, a pronounced stiffness increase could be observed. The reduced mechanical stability at temperatures around 60 °C and higher is a drawback for the FFD method, because the softened material tended to clog the feedstocks' granule hopper located above the extruder due to ascending heat.

Table 4. Extruded PEG(8000)60/PMMA40 filaments (solid load 60 vol%).

Feedstock	Extrusion Temperature (°C)	Filament Diameter (mm)
SA	106	2.81 ± 0.04
PAT77P	70	2.87 ± 0.02
PAT659/CB	n.a.	n.a.

Table 5. FFF and FFD printing parameters for the selected PEG (8000)60/PMMA40 applying both PAT surfactants (solid load 60 vol%).

Printing Parameter	FFF (X350Pro)	FFD (150H)
Temperature (°C)	210	220
Platform temperature (°C)	50	60
Printing speed (mm/s)	5–10	5
Printing speed first layer (mm/s)	3	3
Nozzle diameter (mm)	0.15/0.3/0.4	0.4
Infill (%)	105	105
Overlap contour lines with infill (mm)	0.2	0.2

Both printing methods—FFF and FFD—allowed the printing of small parts in a good and reliable quality, but the surface appearance of the FFD-printed parts was slightly worse than the FFF-printed ones, which can be attributed to the not yet optimized printhead. Figure 15a–c show FFF-printed parts with different demanding geometries, Figure 15d via FFD-printed screws. Due to the huge viscosity in FFF and FFD printing conditions, especially the process related small shear rates, it was not possible to print parts applying a feedstock with 65 vol%.

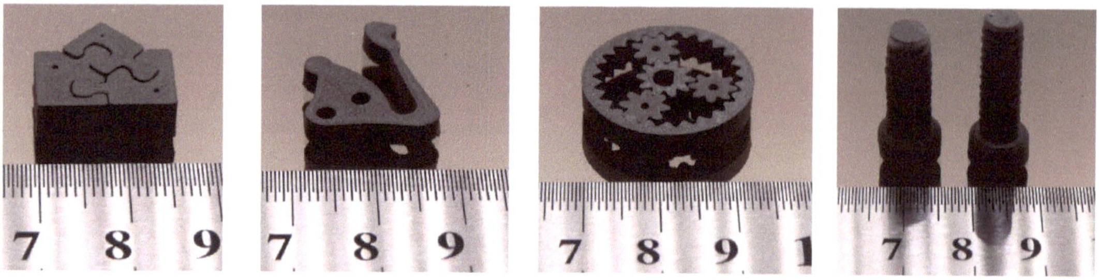

(**a**) (**b**) (**c**) (**d**)

Figure 15. Printed PEG (8000)60/PMMA40 feedstock (solid load 60 vol%). (**a**) Puzzle (FFF); (**b**) test structure with long cantilever and holes [18]; (**c**) planetary gear train (FFF); (**d**) screws (FFD).

3.5. Debinding and Sintering

As in the wax/PE-based feedstocks, debinding was undertaken as a two-step process: the liquid pre-debinding (water, 40 °C, 24 h) was followed by a thermal debinding. Figure 16a shows the PEG removal amount for PEG/PVB with different PEG M_W (SA as surfactant) and different test samples (filament, cuboid, disk; see Figure 1 in [36]). Quite surprisingly, the filament showed the best PEG elimination in all investigated systems irrespective of the PEG M_W. Only in the case of the PEG 20,000 can a comparable removal of the PEG in the filament as well as in the printed specimen be observed. This unusual behaviour was described in the literature [51] and can be attributed to an enhanced thermal instability at elevated temperatures of the low molecular mass PEGs. The printed specimen experienced in contrast to the filament an additional extraordinary heat treatment during printing (extrusion temperature: 90 °C; printing temperature: 230 °C), which caused a PEG decomposition to water-insoluble decomposition products [51]. A further increase of the exposition time (48 h) in water did not show a significant PEG removal improvement. Figure 16b shows the PEG off-take for different printed samples applying the binder PEG(8000)60/PMMA40 with suitable PAT surfactants. In contrast to the PEG/PVB feedstocks, a rapid liquid debinding of all investigated samples and mixtures up to 90% of PEG could be observed. An impact of the huge FFF/FFD printing temperature and resulting PEG decomposition could be found. These results agree with previous investigations [51].

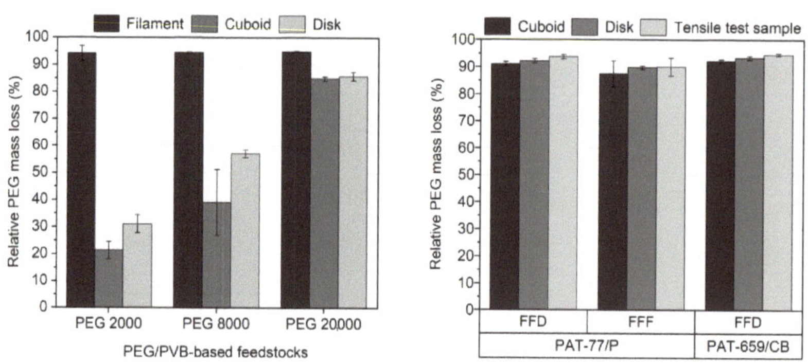

(**a**) (**b**)

Figure 16. Relative PEG mass loss of water-soluble PEG-containing feedstocks after 24 h. (**a**) PEG50/PVB50-based systems; (**b**) PEG (8000)60/PMMA40-based systems with different surfactants and printing method.

Chen et al. also used a similar water-soluble binder system with a Ti6Al4V load of 69.5 vol% [48]. They described a very fast binder removal of almost 100% (sample thickness: 1.5 mm) at 60 °C [48]. Subsequently, after liquid pre-debinding, the samples were thermally debinded followed by a sintering step. Some of the samples were additionally HIPed following ASTM F3001-14 [27] (heating and cooling rate 15 K/min, max. temperature 920 °C for 2 h, max. pressure 100 MPa, argon atmosphere) to achieve higher density values. Figure 17 presents different sintered test structures with various geometric features.

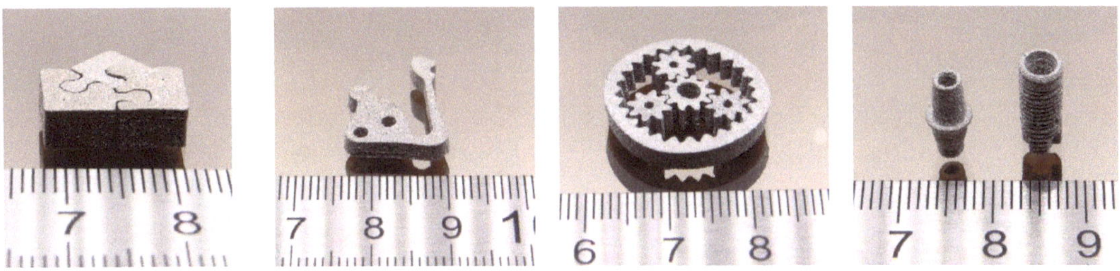

(**a**) (**b**) (**c**) (**d**)

Figure 17. Sintered Ti6Al4V parts (initial solid load: 60 vol%. (**a**) Puzzle (binder PEG/PMMA); (**b**) test structure with long cantilever and holes (binder PEG/PMMA); (**c**) planetary gear train (binder PEG/PVB); (**d**) dental implants (binder PEG/PVB).

3.6. *Sinter Part Characterization*

3.6.1. Density

Being at the end of a process chain, all earlier steps have a pronounced impact on the final density and the resulting mechanical properties. Figure 18a shows the achieved density values for different PEG/PVB mixtures, and Figure 18b for PEG/PMMA systems, both with FFF as the printing method. The red lines in Figure 18 represent the reference value for usage as medical implants with respect to ASTM F2885-17 (without HIP: 96%, with HIP: 98%) [27]. In the case of the PEG/PVB feedstocks, the smaller achievable density of the printed parts in contrast to the filament can be seen again, especially in the case of the PEG 8000 containing mixture (Figure 18a). The usage of PEG 20,000 and the resulting liquid debinding behaviour with the excellent PEG 20,000 removal allowed final density values slightly lower than the value defined by the ASTM F2885-17 value of 96% of the theoretical density. In the case of the PEG/PMMA feedstocks applying the PAT surfactants, the two different nozzle diameters did show a relevant influence on the final sintered part density, because the larger nozzle diameter allowed higher density values between 95% (cuboid) and 96% (disk), which can be attributed to a reduction in the FFF-related void generation during printing (Figure 18b). The additional HIP densification enabled densities higher than the requested 98% up to values close to 100%. It was shown earlier [36] that, e.g., the large infill value of 105% helped to minimize the void generation during printing, allowing huge density values. The use of FFD instead of FFF delivered, in general, lower density values of around 3% percentage points irrespective of which PAT surfactant was used. The same PEG/PMMA feedstock was used in powder injection molding like the wax/PE feedstocks described in [36]. It was possible to achieve densities without HIP around 97% and with HIP close to 100%. These huge values can be explained by the high-pressure injection process and the removal of any voids during cavity mold filling. The achieved density values before and after the HIP process were almost identical to the ones using wax/PE-based feedstocks with a 60 vol% solid Ti6Al4V load as described earlier [36]. Singh et al. [33] reported a best sinter density of 95.6% with an initial solid load of 59 vol% in the applied proprietary binder, which is comparable with the results presented here.

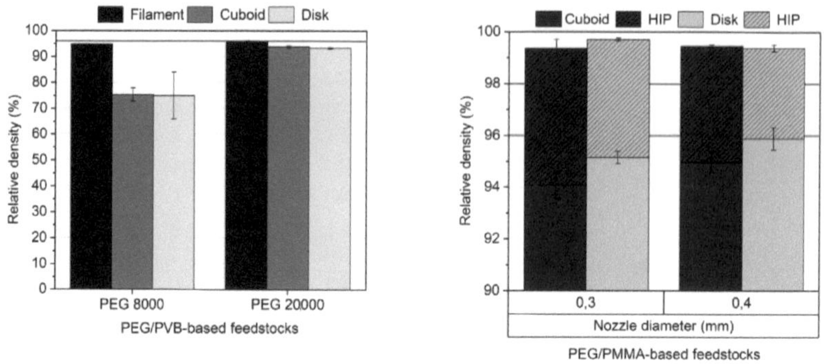

(**a**) (**b**)

Figure 18. Relative sinter density values (initial solid load 60 vol%) of different FFF-printed test samples: (**a**) PEG/PVB-based feedstocks; (**b**) PEG/PMMA-based feedstocks. Additional densification by HIP marked by hatching. Red lines with respect to values described in ASTM F2884-17 [27].

3.6.2. Microstructure

The achievable density depends on the feedstock composition as well as the individual processing steps. Exemplarily, and following the specimen structure described in Figure 1 of [36], Figure 19a,b show the SEM images of a sintered FFF or via PIM fabricated part. The density of the FFF-printed part is 94.7% and the related value for the by PIM fabricated part is 96.9% of the theoretical density, respectively. The number of voids in the case of the FFF-fabricated part slightly increased, which correlates to the lower density value.

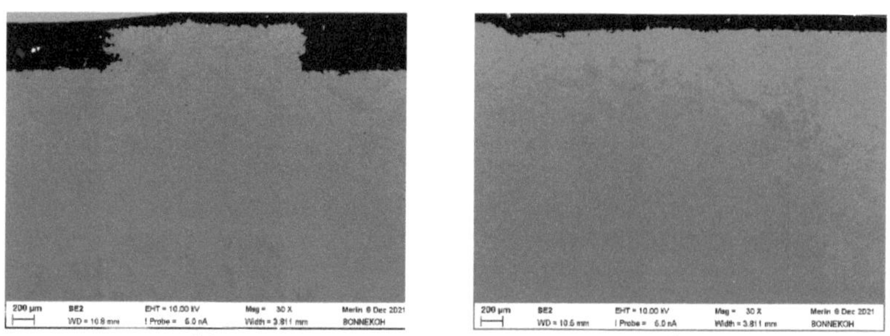

(**a**) (**b**)

Figure 19. SEM fracture images of sintered parts PEG(8000)60/PMMA40 feedstock (solid load: 60 vol%, 5 wt% PAT-77P). (**a**) sintered (FFF); (**b**) sintered (PIM).

3.6.3. Elemental Analysis

The maximum allowed oxygen and carbon content in medical applications is set to 0.2 wt% (oxygen) and 0.08 wt% (carbon) [27,52]. Figure 20a shows for the PEG (8000)60/ PMMA40 feedstock the remaining oxygen and carbon concentration values after sintering and partially additional HIP process for FFF and PIM. For better comparison, the previously obtained data applying wax/PE binders were added [36]. The red lines represent the reference values following ASTM F2885-17 [27}. The sintered FFF part fulfils the ASTM criteria for acceptable oxygen and carbon concentration. The HIPed part and the by PIM fabricated part missed the oxygen concentration limit clearly. The wax/PE feedstock, with the exception of the HIPed part, fulfil the oxygen concentration limit. One has to consider that the PEG/PMMA binder system contains a lot of oxygen atoms in the organic molecules, which may increase the remaining oxygen content after thermal processing even under oxygen-free debinding and sintering conditions. The carbon content is not a critical issue. Figure 20b shows a SEM image of the Ti6Al4V microstructure after etching with ammonium hydrogen fluoride, originated from the PEG/PMMA-based feedstock. The typical combined α/β lamellar texture is present, which is in conformity with the requirements defined in [27,52].

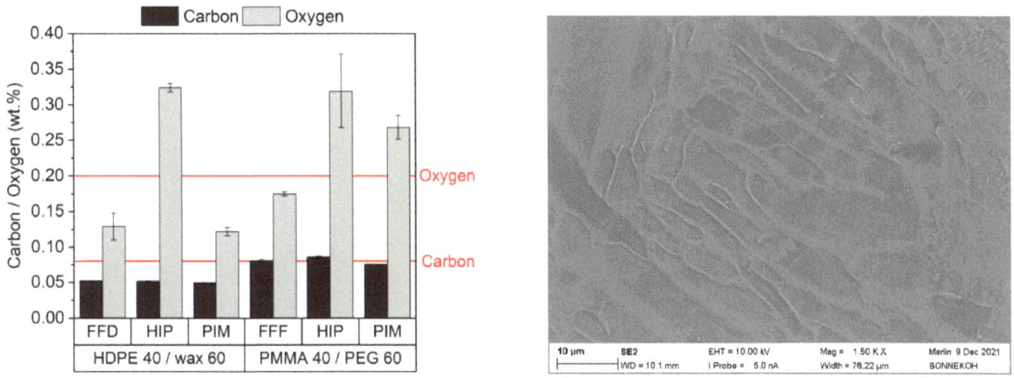

(**a**) (**b**)

Figure 20. (**a**) Elemental analysis after sintering and HIP applying different replication methods. The red line represents the maximum allowed values according to [27,52]. (**b**) SEM image of sintered and additional HIPed microstructure (initial composition: PEG (8000)60/PMMA40, 5 wt% PAT-77P. solid load 60 vol%, FFF printing, density 99.9%).

4. Conclusions and Outlook

It was possible to combine previous work, dealing on the one hand with a partially water-soluble binder system, developed for powder injection molding, and on the other hand the development of FFF and FFD, for the realization of Ti6Al4V parts in good quality almost suitable for use in medical engineering. These water-soluble binders allow an eco-friendly processing by avoiding the harmful hexane in the liquid pre-debinding step of the established wax/PE binder system. A comprehensive rheological investigation on the two binder systems, PEG/PVB and PEG/PMMA, including the change of the PEG´s average molecular weight, enabled the formulation of feedstocks with a solid load of 60 vol%, which were suitable for additive manufacturing of Ti6Al4V parts via FFF and partially for FFD. Liquid pre-debinding in water and thermal debinding as well as sintering metal parts enabled a density close to 96% theoretical density, which is a default value defined by the ASTM F2885-17 for use in medical applications. A further densification by HIP delivered density values better than 99%, fulfilling the ASTM F2885-17.

Future research will focus on the estimation of the mechanical properties as well as on further reduction in the oxygen and carbon content according to ASTM F2885-17.

Author Contributions: Conceptualization, S.A., D.N. and T.H.; methodology, S.A., D.N. and T.H.; validation, R.E., S.A., D.N. and T.H.; formal analysis, R.E., S.A., D.N. and T.H.; investigation, R.E.; resources, T.H.; data curation, R.E.; writing—original draft preparation, R.E. and T.H.; writing—review and editing, T.H.; visualization, R.E. and T.H.; supervision, S.A. and D.N.; project administration, S.A. and D.N.; funding acquisition, T.H. All authors have read and agreed to the published version of the manuscript.

Funding: The APC was funded by the publication fund of the Karlsruhe Institute of Technology.

Institutional Review Board Statement: Not applicable.

Informed Consent Statement: Not applicable.

Data Availability Statement: Not applicable.

Acknowledgments: The authors thank their colleagues A. Klein for heat treatment, D. Bohlich for metallography, C. Bonnekoh for SEM image recording, T. Bergfeldt for the C, O analysis and M. Offermann for powder characterization. In addition, the support from W. Limberg at Helmholtz-Center Hereon for performing the first sinter experiments is gratefully acknowledged. The authors also thank E.& P. Wuertz GmbH, Bingen, Germany, for the kind donation of the two PAT surfactants.

Conflicts of Interest: The authors declare no conflict of interest.

References

1. Kluck, S.; Hambitzer, L.; Luitz, M.; Mader, M.; Sanjaya, M.; Balster, A.; Milich, M.; Greiner, C.; Kotz-Helmer, F.; Rapp, B.E. Replicative manufacturing of metal moulds for low surface roughness polymer replication. *Nat. Commun.* **2022**, *13*, 5048. [CrossRef] [PubMed]
2. Gohn, A.M.; Brown, D.; Mendis, G.; Forster, S.; Rudd, N.; Giles, M. Mold inserts for injection molding prototype applications fabricated via material extrusion additive manufacturing. *Addit. Manuf.* **2022**, *51*, 102595. [CrossRef]
3. Sarabia-Vallejos, M.A.; Rodríguez-Umanzor, F.E.; González-Henríquez, C.M.; Rodríguez-Hernández, J. Innovation in Additive Manufacturing Using Polymers: A Survey on the Technological and Material Developments. *Polymers* **2022**, *14*, 1351. [CrossRef]
4. Ranjan, R.; Kumar, D.; Kundu, M.; Chandra Moi, S. A critical review on Classification of materials used in 3D printing process. *Mater. Today Proc.* **2022**, *61*, 43–49. [CrossRef]
5. Cramer, C.L.; Ionescu, E.; Graczyk-Zajac, M.; Nelson, A.T.; Katoh, Y.; Haslam, J.J.; Wondraczek, L.; Aguirre, T.G.; LeBlanc, S.; Wang, H.; et al. Additive manufacturing of ceramic materials for energy applications: Road map and opportunities. *J. Eur. Ceram. Soc.* **2022**, *42*, 3049–3088. [CrossRef]
6. Blakey-Milner, B.; Gradl, P.; Snedden, G.; Brooks, M.; Pitot, J.; Lopez, E.; Leary, M.; Berto, F.; du Plessis, A. Metal additive manufacturing in aerospace: A review. *Mater. Des.* **2021**, *209*, 110008. [CrossRef]
7. Saleh Alghamdi, S.; John, S.; Roy Choudhury, N.; Dutta, N.K. Additive Manufacturing of Polymer Materials: Progress, Promise and Challenges. *Polymers* **2021**, *13*, 753. [CrossRef]
8. Lakhdar, Y.; Tuck, C.; Binner, J.; Terry, A.; Goodridge, R. Additive manufacturing of advanced ceramic materials. *Prog. Mater. Sci.* **2021**, *116*, 100736. [CrossRef]
9. Piotter, V.; Hanemann, T.; Heldele, R.; Mueller, T.; Plewa, K.; Ruh, A. Metal and Ceramic Parts fabricated by Microminiature powder injection molding. *Int. J. Powder Metall.* **2010**, *43*, 21–28.
10. Clemens, F.J.; Kerber, A. FDM/FFF an Alternative to CIM Manufacturing of Prototype and Small Quantities of Ceramic Part? *Ceram Appl.* **2020**, *8*, 27–31.
11. Noetzel, D.; Eickhoff, R.; Hanemann, T. Fused Filament Fabrication of Small Ceramic Components. *Materials* **2018**, *11*, 1463. [CrossRef]
12. Noetzel, D.; Hanemann, T. New Feedstock System for Fused Filament Fabrication of Sintered Alumina Parts. *Materials* **2020**, *13*, 4461. [CrossRef] [PubMed]
13. Eickenscheidt, M.; Langenmair, M.; Dbouk, A.; Noetzel, D.; Hanemann, T.; Stieglitz, T. 3D-Printed Hermetic Alumina Housings. *Materials* **2021**, *14*, 200. [CrossRef]
14. Cano, S.; Gonzalez-Gutierrez, J.; Sapkota, J.; Spoerk, M.; Arbeiter, F.; Schuschnigg, S.; Holzer, C.; Kukla, C. Additive manufacturing of zirconia parts by fused filament fabrication and solvent debinding: Selection of binder formulation. *Addit. Manuf.* **2019**, *26*, 117–128. [CrossRef]
15. Kataoka, Y.; Yokota, K. Fabrication of Three-dimensional Zirconia Ceramics by Material Extrusion. *J. Soc. Powder Technol. Jpn.* **2020**, *57*, 520–525. [CrossRef]
16. He, Q.; Jiang, J.; Yang, X.; Zhang, L.; Zhou, Z.; Zhong, Y.; Shen, Z. Additive manufacturing of dense zirconia ceramics by fused deposition modeling via screw extrusion. *J. Eur. Ceram. Soc.* **2021**, *41*, 1033–1040. [CrossRef]
17. Hadian, A.; Koch, L.; Koberg, P.; Sarraf, F.; Liersch, A.; Sebastian, T.; Clemens, F. Material Extrusion Based Additive Manufacturing of Large Zirconia Structures Using Filaments with Ethylene Vinyl Acetate Based Binder Composition. *Addit. Manuf.* **2021**, *47*, 102227. [CrossRef]
18. Noetzel, D.; Eickhoff, E.; Pfeifer, C.; Hanemann, T. Printing of Zirconia Parts via Fused Filament Fabrication. *Materials* **2021**, *14*, 5467. [CrossRef]
19. Kukla, C.; Cano, S.; Moritz, T.; Müller-Köhn, A.; Courtney, P.; Hampel, S.; Holzer, C. Multimaterial Components by Material Extrusion-Fused Filament Fabrication (ME-FFF)—Production of an Infrared Heater. *Ceram. Forum Int.* **2019**, *96*, 22–27.
20. Lieberwirth, C.; Harder, A.; Seitz, H. Extrusion Based Additive Manufacturing of Metal Parts. *J. Mech. Eng. Autom.* **2017**, *7*, 79–83. [CrossRef]
21. Masood, S.H.; Song, W.Q. Development of new metal/polymer materials for rapid tooling using Fused deposition modelling. *Mater. Des.* **2004**, *25*, 587–594. [CrossRef]
22. Gonzalez-Gutierrez, J.; Cano, S.; Schuschnigg, S.; Kukla, C.; Sapkota, J.; Holzer, C. Additive Manufacturing of Metallic and Ceramic Components by the Material Extrusion of Highly-Filled Polymers: A Review and Future Perspectives. *Materials* **2018**, *11*, 840. [CrossRef]

23. Singh, G.; Missiaen, J.-M.; Bouvard, D.; Chaix, J.-M. Copper extrusion 3D printing using metal injection moulding feedstock: Analysis of process parameters for green density and surface roughness optimization. *Addit. Manuf.* **2021**, *38*, 101778. [CrossRef]
24. Kukla, C.; Gonzalez-Gutierrez, J.; Burkhardt, C.; Holzer, C. The Production of Magnets by FFF-Fused Filament Fabrication. In Proceedings of the Euro PM2017, Milano, Italy, 1–5 October 2017.
25. Parenti, P.; Puccio, D.; Colosimo, B.M.; Semeraro, Q. A new solution for assessing the printability of 17-4 PH gyroids produced via extrusion-based metal AM. *J. Manuf. Process.* **2022**, *74*, 557–572. [CrossRef]
26. Riaz, A.; Töllner, P.; Ahrend, A.; Springer, A.; Milkereit, B.; Seitz, H. Optimization of composite extrusion modeling process parameters for 3D printing of low-alloy steel AISI 8740 using metal injection moulding feedstock. *Mater. Des.* **2022**, *219*, 110814. [CrossRef]
27. *ASTM F2885-17*; Standard Specification for Metal Injection Molded Titanium-6Aluminum-4Vanadium Components for Surgical Implant Applications. American Society for Testing and Materials International: West Conshohocken, PA, USA, 2017.
28. Zhang, G.; Lu, X.; Li, J.; Chen, J.; Lin, X.; Wang, M.; Tan, H.; Huang, W. In-situ grain structure control in directed energy deposition of Ti6Al4V. *Addit. Manuf.* **2022**, *55*, 102865. [CrossRef]
29. Li, H.; Yu, Y.; Li, Y.; Lin, F. Effects of the higher accelerating voltage on electron beam powder-bed based additive manufacturing of Ti6Al4V alloy. *Addit. Manuf.* **2022**, *50*, 102579. [CrossRef]
30. Damon, J.; Czink, S.; Schüßler, P.; Antusch, S.; Klein, A.; Send, S.; Dapprich, D.; Dietrich, S.; Schulze, V. Mechanical surface treatment of EBM Ti6Al4V components: Effects of the resulting surface layer state on fatigue mechanisms and service life. *Mater. Sci. Eng. A* **2022**, *849*, 143422. [CrossRef]
31. Zhang, F.; Wang, K.; Li, Y.; Chen, Y.; Yang, M.; Wang, M.; Clare, A.T. Composition fine-tuning for directed energy deposition of Ti-6Al-4V. *J. Mater. Process. Technol.* **2022**, *299*, 117321. [CrossRef]
32. Thompson, Y.; Polzer, M.; Gonzalez-Gutierrez, J.; Kasian, O.; Heckl, J.P.; Dalbauer, V.; Kukla, C.; Felfer, P.J. Fused Filament Fabrication-Based Additive Manufacturing of Commercially Pure Titanium. *Adv. Eng. Mater.* **2021**, *23*, 2100380. [CrossRef]
33. Singh, P.; Balla, V.K.; Atre, S.V.; German, R.M.; Kate, K.H. Factors affecting properties of Ti-6Al-4V alloy additive manufactured by metal fused filament fabrication. *Powder Technol.* **2021**, *386*, 9–19. [CrossRef]
34. Singh, P.; Balla, V.K.; Tofangchi, A.; Atre, S.V.; Kate, K.H. Printability studies of Ti-6Al-4V by metal fused filament fabrication (MF3). *Int. J. Refract. Met. Hard Mater.* **2020**, *91*, 105249. [CrossRef]
35. Gloeckle, C.; Konkol, T.; Jacobs, O.; Limberg, W.; Ebel, T.; Handge, U.A. Processing of Highly Filled Polymer-Metal Feedstocks for Fused Filament Fabrication and the Production of Metallic Implants. *Materials* **2020**, *13*, 4413. [CrossRef]
36. Eickhoff, R.; Antusch, S.; Baumgärtner, S.; Nötzel, D.; Hanemann, T. Feedstock Development for Material Extrusion-Based Printing of Ti6Al4V Parts. *Materials* **2022**, *15*, 6442. [CrossRef]
37. Bek, M.; Gonzalez-Gutierrez, J.; Kukla, C.; Pušnik Črešnar, K.; Maroh, B.; Slemenik Perše, L. Rheological Behaviour of Highly Filled Materials for Injection Moulding and Additive Manufacturing: Effect of Particle Material and Loading. *Appl. Sci.* **2020**, *10*, 7993. [CrossRef]
38. Mezger, T.G. *The Rheology Handbook*, 4th ed.; Vincentz Network: Hanover, Germany, 2014; ISBN 3-86630-650-4.
39. Basir, A.; Sulong, A.B.; Jamadon, N.H.; Muhamad, N.; Emeka, U.B. Process Parameters Used in Macro/Micro Powder Injection Molding: An Overview. *Met. Mater. Int.* **2020**, *27*, 2023. [CrossRef]
40. Metal Powder Ti6Al4V, Data Sheet, Heraeus. Available online: https://www.heraeus.com/media/media/group/doc_group/products_1/additivemanufacturing/datasheets_en/Ti6Al4V.pdf (accessed on 18 July 2022).
41. Medesi, A.J.; Noetzel, D.; Hanemann, T. PVB/PEG-Based Feedstocks for Injection Molding of Alumina Microreactor Components. *Materials* **2019**, *12*, 1219. [CrossRef]
42. Medesi, A.J.; Noetzel, D.; Wohlgemuth, J.; Franzreb, M.; Hanemann, T. Ceramic Injection Moulding using 3D-Printed Mould Inserts. *Ceram. Mod. Technol.* **2019**, *1*, 104–110. [CrossRef]
43. Weber, O.; Hanemann, T. Green-conscious ceramic injection moulding. In *Advanced Processing and Manufacturing Technologies for Structural and Multifunctional Materials VI*; Ihji, T., Singh, M., Halbig, M., Mathur, S., Eds.; John Wiley & Sons: Hoboken, NJ, USA, 2013; pp. 63–71.
44. Weber, O.; Hanemann, T. Ceramic injection moulding using a partially water-soluble binder system: Effect of back-bone polymers on the process. In *Advanced Processing and Manufacturing Technologies for Structural and Multifunctional Materials VI*; Ihji, T., Singh, M., Halbig, M., Mathur, S., Eds.; John Wiley & Sons: Hoboken, NJ, USA, 2013; pp. 52–61.
45. Hanemann, T.; Heldele, R. Modern Alchemy: Ceramic Feedstock Optimization by Surfactant Screening. *CFI/Ber. DKG* **2010**, *87*, E38–E40.
46. Hanemann, T.; Weber, O. Thermoplastic PMMA/PEG binder system for micro ceramic injection moulding. In Proceedings of the 10th International Conference on Multi-Material Micro Manufacture, San Sebastian, Spain, 8–10 October 2013; pp. 231–234.
47. Chuankrerkkul, N.; Messer, P.F.; Davies, H.A. Flow and void formation in powder injection moulding feedstocks made with PEG/PMMA binders Part 1—Experimental observations. *Powder Metall.* **2013**, *51*, 66–71. [CrossRef]
48. Chen, G.; Cao, P.; Wen, G.; Edmonds, N. Debinding behaviour of a water soluble PEG/PMMA binder for Ti metal injection moulding. *Mater. Chem. Phys.* **2013**, *139*, 557–565. [CrossRef]
49. Hayat, M.D.; Wen, G.; Zulkifli, M.F.; Cao, P. Effect of PEG molecular weight on rheological properties of Ti-MIM feedstocks and water debinding behaviour. *Powder Technol.* **2015**, *270*, 293–301. [CrossRef]
50. Available online: https://spool-database.info/pvb-filament-was-ist-das (accessed on 1 February 2023).

51. Weber, O. Wasserlösliche Bindersysteme zum Minimieren von Pulver-Binder-Segregationseffekten im Mikropulverspritzguss. Ph.D. Thesis, University of Freiburg, Freiburg, Germany, 2015.
52. *DIN EN ISO 5832-3:2020-11*; Implants for Surgery—Metallic Materials—Part 3: Wrought Titanium 6-Aluminium 4-Vanadium Alloy (ISO/DIS 5832-3:2020-11). DIN: Berlin, Germany, 2020.

MDPI AG
Grosspeteranlage 5
4052 Basel
Switzerland
Tel.: +41 61 683 77 34

Materials Editorial Office
E-mail: materials@mdpi.com
www.mdpi.com/journal/materials

www.ingramcontent.com/pod-product-compliance
Lightning Source LLC
Chambersburg PA
CBHW041349210926
51850CB00013B/3033
* 9 7 8 3 7 2 5 8 3 2 5 6 9 *